The Truth in the Mirror

How a Global Business Leader Became an Expert in Subtle Energy and Animal Telepathy

A groundbreaking true story of intuition, normalising psychic senses, and the Heather Ogilvie Theory of Life Force Transmission. Redefining healing, leadership, and the future of human intelligence.

Heather Ogilvie

The Truth in the Mirror

ISBN 978-1-7394347-9-3

Publisher information: Ogilvie Advisory Ltd T/A The Islay Wellness Academy

Disclaimer: This book is intended for informational and inspirational purposes only. It does not constitute medical, psychological, or professional health advice. Always consult with a qualified healthcare provider before making changes to your health or well-being practices.

The Truth in the Mirror

Contents

The Truth in the Mirror

Prologue: A Dog Saved My Life (and I Saved His)

Askar's diagnosis and surgery through animal telepathy

This is the Story that Changed Everything.

I start this narrative with my dog, Askar.

He was getting older, but not in a way that worried me. He was fit, bright, engaged, and full of life. And then, slowly, things started to shift. He stopped eating with his usual enthusiasm. He shook his head strangely sometimes, like something was jolting him. There were moments, especially when he drank cold water, where he froze, as if in pain. His face didn't look right. I knew something was wrong.

The vets ran their tests. They found a bit of leftover root from a previous tooth extraction. He had surgery to remove it, and it helped, but... not enough. That deep discomfort was still there, and I could feel it, like a drumbeat in the background. But nobody could tell me what it was.

That's when someone mentioned an animal communicator.

Let me be clear: I was not into this kind of thing. I didn't even believe in "this kind of thing." I come from a world of science, logic, and proof. Raised by atheists, trained in strategy.

I've run companies across the globe, made decisions based on numbers and evidence. The only energy I paid attention to was the vibe in a boardroom.

But something in me was open. Just a crack.

So I sent a photo to a woman called Julie. She was in Utah. I was in Scotland. She asked for nothing but the image, no backstory, no health notes, no cues. Just his eyes, and mine.

The Truth in the Mirror

The session we had still lives in me like a marker in time. She said a lot of things, some general, some surprisingly accurate. But then she said, "*He's got pain in the right side of his face. Something's swollen in his nostril. He can't smell properly. There's a blockage in his sinus.*"

And everything in me stilled. That was the piece I couldn't name but had been feeling all along.

A few days later, I was at a physio appointment for Askar's leg, which I nearly cancelled due to torrential rain. The physio had a thermal imaging camera. I asked her to scan his face. She did, and there it was. A cold, swollen pocket around his right sinus. The inflammation was real. Visible. Undeniable.

Armed with that image, we went back to the vets. Further scans. Deeper investigation. A rare fracture in a hidden tooth root. A severe infection had spread from there into his cheekbone and sinus. If we hadn't found it, it would have kept spreading. Quietly. Invisibly. Terminally.

The surgery that followed saved his life.

That moment, right there, was the moment everything changed for me.

Because once you experience something that *shouldn't* be possible — but is — you can't go back. You can't pretend it didn't happen. You can't explain it away.

This book is not about becoming mystical or magical. It's not about abandoning logic. It's about expanding what we include in our definition of reality.

What happened with Askar wasn't faith — it was evidence. Real, grounded, verifiable information of animal-to-human telepathy delivered through non-verbal means. That was the door.

And once it opened, my entire life changed, because I chose to step through.

The Truth in the Mirror

Here is the actual thermal image of Askar's face taken by the physiotherapist, looking down onto his face, eye sockets are the bright white spots on the left, his nose is dark on the right, and the arrow points to the blockage issue in his right-hand sinus area, and you can see the swelling extending down his nose.

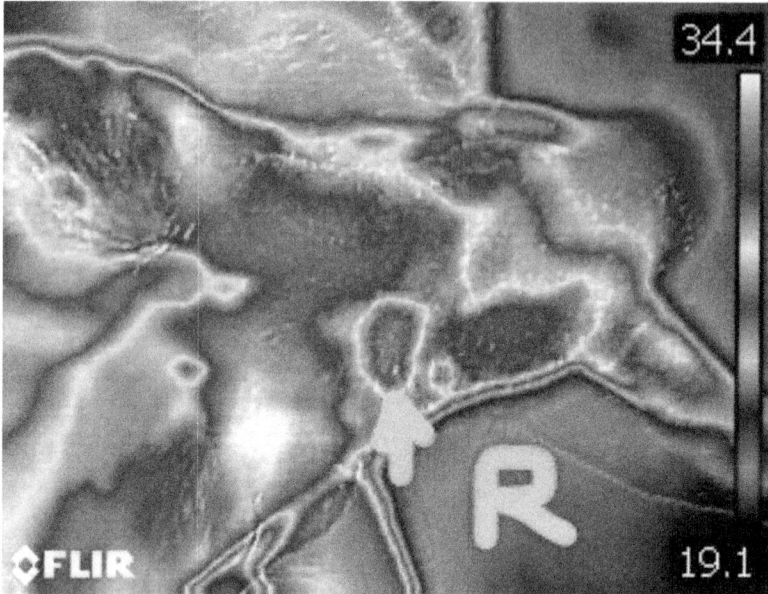

Author's note: It has taken me a long time to write this book. I had to overcome a lot of internal bias and recalibration of my own hardwired beliefs, as well as my ego's fears of exposure. However, this narrative has an importance that is far greater than mine. The implications extend beyond our current human narrative, and the long-term potential impact and avenues of research have the potential to benefit many future generations of humans and animals on our beautiful planet.

Part One: The Turning Point

How a Corporate Strategist Stumbled into the Future of Human Perception

The Truth in the Mirror

Chapter 1: The Shock of Evidence

How one dog, one communicator, and one moment changed everything.

The Full Askar Story

I was raised in a house that revered science and logic; atheism wasn't even a debate; it was simply the framework we all lived by. Questions like "Do you believe in life after death?" weren't asked. They didn't need to be. The answer was always: no. And I was fine with that. Curious, yes, but rational, grounded, and focused on facts. Even as I grew older and built a career in corporate strategy and business transformation, I stayed that way: evidence-led, practical, logical.

So when I tell you that the event that changed everything began with a woman in Utah and a photograph of my dog, I need you to understand just how hard this was for me to believe, even as it was happening.

His name was Askar. Stoic, gentle, loyal to the core. He was the kind of dog who lit up every room with quiet dignity. The kind of dog who didn't complain, even when he was in pain. That summer, he'd been limping on and off, and something wasn't right with his face. He had "brain freeze" moments drinking cold water, and his eating had slowed. The vet had already performed a tooth extraction and a lengthy operation to remove some remaining root. That helped, but something still wasn't right. And his leg? No clear diagnosis.

Around the same time, I was on a book-writing call and a friend, Karen, casually mentioned a woman called Julie, an animal communicator who telepathically received messages from pets. I was intrigued, skeptical, but curious enough to ask questions. Julie worked with just a photo of your pet. No background. Just your eyes and theirs.

The Truth in the Mirror

Honestly? I didn't think anything would come of it. But I booked a session anyway. I sent her a photo of Askar and me, then followed her instruction to say nothing else. No clues. No backstory. And when we got on the call, Askar was next to me, awake, just as she'd requested.

Then the floodgates opened.

She said he had a sore right front leg, above the elbow. That was exactly where his limp had been coming from. She said he'd hurt it digging in something soft and wet, she asked if it could've been mud. I said, "Sand?" She said yes. That would explain it. He didn't mind, she said, he'd been having the best day ever.

He had. He'd been surfing, digging, playing, living his best life. That was the day he stepped into a sand hole and wrenched his shoulder.

Then she asked, "Who's the big golden Labrador-type dog? Heavyset, big bones, gentle. Askar says he's not the brightest, but very sweet." That was Merlin, my parents' dog. Julie described how Askar felt he had to help Merlin understand things, but loved him all the same.

Julie described many other moments, events with myself and Askar, with my other old dog who was no longer with us, and specific behaviours. There was no possible way she could have known the level of information she provided, no public source, such as any of my Facebook posts about my dogs, could ever have informed her.

But it wasn't until the very end of the call that Julie said something that changed everything:
 "*Askar says he has pain on the right-hand side of his face. He can't smell. His sinus is blocked.*"

Julie told me I needed to take him to the vets and get it checked out.

Something inside me clicked. It was like a massive internal shift. I felt the truth of it before I understood the science. I knew this wasn't just a random guess.

The Truth in the Mirror

I discussed the possibility of a sinus issue with the vets, but they were uncertain that even an X-ray would detect anything, as it was a soft tissue area. I couldn't possibly put Askar through that without more evidence, and a phone call with a lady in Utah who had never met my dog was certainly not proof enough!

A few days later, we had a physio appointment for Askar's leg injury. Torrential rain, flash flooding, I almost didn't go. But I did. And that physio had a thermal imaging camera.

I asked if anything in the sinus area would show up on the camera. She said yes. She snapped the image and — there it was. A cold spot. A visible area of inflammation in the exact place Julie had mentioned.

We took that image to the vet. A CT scan followed. The result? A fractured tooth root. Infection spreading into the cheekbone, the sinus, and the eye socket.

Julie, this woman on the other side of the world, had described it to me without ever laying eyes on him in person.

And that was the moment the door to this new world creaked open.

Atheist and Scientific Upbringing

I didn't grow up in a world where anything like this was allowed to be true.

My childhood was shaped by logic, science, and proof. My parents were atheists, staunchly so, and I was raised to question everything, to lean on facts, and to be deeply suspicious of anything that couldn't be measured or repeated. There were no angels, no spirits, no God in our house. We didn't "feel" things, we explained them.

It wasn't cruel, nor uncomfortable. It was just the framework. Everything had to make sense, and if it didn't, it got dismissed, and I was totally happy with this for most of my life.

The Truth in the Mirror

That mindset served me well in business. I've led teams across the globe, turned around companies on the edge of collapse, negotiated with hard-nosed boards and built strategy off the back of spreadsheets. I know what it is to work in the real world, and to win in it. Logic works.

But it's not the whole picture.

I didn't have a spiritual awakening in childhood. I wasn't "born psychic." I didn't see dead people or talk to animals when I was small. In fact, if you'd told me ten years ago that I'd be writing a book about psychic intelligence, subtle energy, and telepathic communication, I would have stared at you in horror.

And yet, here I am.

It's not that the science is wrong. It is just immeasurably incomplete. I've come to realise that atheism, for all its clarity, is just as much a belief system as any religion. It's still a lens, a structure. But the deeper I've gone into this work, the more I've had to pull that lens off, painfully, sometimes, and admit that reality is far more layered than I was taught.

I still don't subscribe to religion. But I do believe in intelligence. And what I've found, time and time again, is that there is a substantial possibility that every atom is intelligent. Every moment holds data. And when we learn how to listen, the world gets louder and clearer.

Disbelief, Then Proof — Reality Breaks Through

I didn't disbelieve Julie. That's the strange thing.

What she said landed. It felt real. Specific. Resonant. It matched my dog, and it matched things she couldn't possibly have known. But it also pushed hard against everything I'd ever been taught about how the world works.

Telepathy? Across the Atlantic? With a dog?

The Truth in the Mirror

My logical brain couldn't find a place to put it. So I didn't reject it — but I didn't lean all the way in, either. I let it percolate. I focused on the immediate: Askar's health, the next steps, getting the right kind of help. There wasn't time to process a paradigm shift.

And then came the thermal imaging.

That was the moment it all landed. No room for ambiguity, no space for doubt. The exact area she had described was lit up on the scan. Cold, inflamed, visibly swollen. Follow-up scans confirmed it — an infected root fracture, buried deep in his face, unseeable by the usual checks. The surgery was complex but successful. Without it, things would have deteriorated quickly.

Julie hadn't given me a belief. She'd given me information. Real, accurate, useful information that changed the outcome of my dog's life.

And that's when the shift began.

Not into some mystical worldview. Not into airy spirituality. But into a quiet, unsettling realisation: we have underestimated something very important. Something real, measurable, and present — just not fully understood.

Telepathy was no longer a theory. It had moved into the category of "things I've seen for myself."

And once that line is crossed, you can't unsee it.

First Realisation That Something Deeper Is at Work

After Askar's recovery, I didn't dive straight into animal communication. I wasn't "woo-woo", I wasn't even particularly open. I was stunned and quietly trying to make sense of what had just happened. Julie's reading had been so accurate, so specific, that it left me with only one possible conclusion: telepathy had to be real. And that reality, so foreign to my upbringing, needed time to settle.

The Truth in the Mirror

So I let it sit. I focused on the operation, on getting Askar well again, on what I could see and do and manage in the physical world. But something had shifted. I started to watch more carefully. Listen a little closer. Track what I could feel. Because clearly, something else was going on.

And along came the pennies.

At that time, I was managing a large-scale global project, travelling weekly, often exhausted, sometimes numb. My self-worth was in tatters. I liked my team, I liked the work, but something inside me was quietly breaking. I started walking each day, to think, to breathe, to keep the wheels turning. And that's when I found the first one.

Just a penny, lying on the ground.

"See a penny, pick it up, all day long you'll have good luck," I heard myself mutter. So I picked it up.

The next day, another. And the next. And the next. Four days in a row. Four pennies in my pocket.

The following week, I was staying in a different hotel. No pennies for three days — until I had to change rooms. And under the bed? A chocolate coin wrapped in gold foil. I threw out the chocolate, cleaned the wrapper, and used it to wrap the pennies so they wouldn't jingle.

But then — nothing.

And standing in a café a few days later, I suddenly had the thought: am I constraining the flow by wrapping these up? That I was literally symbolising my own restriction. So I unwrapped them, threw away the foil, and as I looked up, there it was. Another penny, lying on the ground outside the café door.

That's when it started to become clear: I only ever found money when I was thinking about my value.

The Truth in the Mirror

Week after week, it continued. I'd find coins whenever I was contemplating how to move forward, how to be more of myself, how to step into my own worth. Always small. Always precise. And always when the question of "Am I enough?" was hanging in the air.

I began talking to the universe, half-jokingly, half-desperately.

"*All right then,*" I said one day, "*if you really exist, show me two sets of money on this trip to London.*"

I found both.

Next, I asked for a *pile of money.* And outside a bookshop on The Strand, I came across a full scatter of loose change — at least 20 coins — spilled like an offering. I gathered them, shared some with people who needed them more than me, and dropped the rest into donation pots at the airport.

Over four months, I found around 200 coins.

I kept testing it. "*Three £1 coins, please,*" I said once. And there they were, under a seat at the airport — three pound coins, no more, no less. Another time I said, "*Outside my house.*" That was too vague, apparently, because one appeared on my birthday, at the crossroads by my home. Not close enough. So I said, "*Right at the gate.*" And a few weeks later? Two two-pence pieces, one on either side of the gate. Exactly as requested.

And then came the moment with Tober.

Tober was my older dog, Askar's brother. The day he passed away was still, sunny, and silent. But as he crossed, two gusts of wind blew through the car, and then it fell calm again. I felt his presence so strongly it filled the car. Later that day, exhausted and raw, I whispered, "*Tober, if you're still here, show me. Give me some money on the ground.*"

Twenty paces later: a 20p coin.

The Truth in the Mirror

That's when I stopped calling this coincidence.

The pennies weren't just about money. They were about presence. Patterns. Participation. They were a way for something intelligent — whatever you want to call it, the universe, your guides, life force — to say, *"I'm here. And I'm listening."*

These were the "wow" moments. They were the foundation. They rebuilt me. They reminded me to pay attention. They showed me how little I had been valuing myself — and how much more was possible when I finally did.

Researching Energy – Finding the Blue Rose

After Askar's operation and recovery, I didn't go chasing mystery. I didn't suddenly become "spiritual." I stayed anchored in what I knew — observation, evidence, lived experience. But I couldn't deny what had happened. Something real had occurred, and I needed to understand it. So I began quietly researching energy, not as a belief, but as a system. A way information might move without wires, screens, or speech. How someone across the Atlantic could describe pain inside my dog's face, in detail, without ever meeting him.

And then came the blue rose.

I was reading a book about signs from the other side, subtle, personal ways the dead might reach the living. It wasn't something I fully believed. But I was open. Curious. Willing to consider. One passage described a blue rose as a sign for a family who had lost a parent, something so unlikely, so out of place, that if you saw one, you'd know.

The next day, I was out walking Askar in the forest.

We weren't on a usual route. There were many paths through those woods, and the one I chose that day wasn't typical for us. The odds of taking that exact trail were low. And yet, in the middle of the path, there it was.

A wooden blue rose.

The Truth in the Mirror

No context. No reason. No flowers had ever been placed on that trail before. A graveyard was half a mile away, and the idea that a dog had somehow carried it from there seemed like the only vaguely rational explanation I could reach for. But still — it didn't make sense.

I didn't pick it up.

I didn't have my phone on me to take a photo. And truthfully, I didn't want to. The whole thing felt too precise, too orchestrated. I rationalised it, filed it away, and walked on — but the discomfort lingered. Something about it felt too weird.

I stored the moment like a seed in my mind. Unprocessed. Quiet. But something had been set in motion.

A day after reading about blue roses, I had walked into one in the most implausible of places. And though I hadn't known what to make of it, I couldn't deny it had happened. It marked the beginning of a slow internal shift — away from dismissing, and towards asking: "What *if this is real?*"

Bumping into a Former Colleague – Signs, Synchronicities, and Something Else

I had no framework for what began happening next. No spiritual lens. No belief system to fall back on. Just the repeated, almost impossible unfolding of events that I couldn't ignore.

It started with a dream after two intensive days of Reiki training. The dream was a vivid one. I dreamed about a person I used to work with, a person who had not treated me well, someone whose presence in my life had left a residue of unfinished business. And the next day, I walked straight into them.

Literally.

The Truth in the Mirror

It was six in the morning at Edinburgh Airport. I had left my house at 4 a.m. and had to turn back to collect something I'd forgotten. I parked, then had to move the car due to unexpected works. Security was delayed. A couple sat next to me and argued loudly, prompting me to get up and move again. All of these seemingly minor events stacked up — and placed me directly, perfectly, on a collision course. When I stepped onto the main concourse, the airport was already bustling with thousands of travellers. Yet, somehow, there were no other people between us. I turned, and there he was, walking directly toward me. Not before, not after — exactly then. That moment.

I walked away in disbelief.

It didn't stop there.

Another time, I was driving alone — 200 miles along a remote Scottish highway — and I was thinking about that airport encounter. As I rounded a corner, someone emerged from a forestry track on foot. It was him. The same person. Travelling at 60 miles per hour, even one or two seconds later, I would never have seen him. He didn't see me, but I saw him clearly as I passed.

Later, while camping with friends, a clear image flashed into my mind — his vehicle passing me on a specific stretch of road. And hours later, that's exactly what happened. Again, no contact, no interaction, but there he was, driving past at the very place I had visualised it.

Still, I tried to rationalise.

But then came the day I said to the universe: "*If he's meant to be in my life, show me. Prove it.*" I travelled to London, deliberately tracing the streets near our old workplace. I booked meetings, I showed up, half-challenging, half-hoping nothing would happen. Four days passed. Nothing. I got on my flight home relieved, certain I could finally release it all. I'd done the test. I could let it go.

The Truth in the Mirror

And then I walked out of the arrivals hall at Edinburgh Airport. And there he was directly in front of me. Standing there, waiting for someone else. In that exact moment. Again.

You cannot logic your way through moments like this. You can try. I did. But something else was clearly in play, something intelligent, unseen, and very precise.

These occurrences have been ongoing for seven years. They're not frequent yet extraordinary in their timing and logistics. Too improbable to explain away. And the hardest part? This person has never seen nor valued me appropriately, either professionally or personally, and I wish they would stop showing up in my life without good reason. Yet, I can't deny what's happening. A subtle energy exchange is in motion. Something is at work beneath the surface.

Has this been a long lesson? A sequence of signs? A reflection of my own frequency? I still don't know. But whatever it is, it has opened something in me that could never be closed again. To this day, I can't fully explain why I've crossed paths with this person so many times, through dreams, visions, and synchronistic implausible real-life encounters. Perhaps our souls were once aligned. Perhaps these moments are part of a much larger message, even bigger than the pennies that urged me to value myself more. But one thing is certain: these experiences played a vital role in awakening me to the undeniable presence of subtle energy.

Even more profound were the times I sensed or visualised something unusual before it happened. Once, I awoke with a deep knowing that I must avoid the woods. These were familiar tracks I'd walked with my dog for over twenty years, places that had never felt unsafe. But for two full days, I felt a powerful aversion to going there. On the third day, news rippled through the local community: a dog walker had been killed in a wooded area just five miles away. The perpetrator was caught. And in that moment, I felt the energy shift — I knew the woods were safe again. That experience, I later learned, was an expression of *precognition*, coupled with *clairsentience*.

The Truth in the Mirror

On another occasion, I was walking past a local café when a strong sensation came over me, accompanied by a clear visual image. I saw two people, a couple, sitting there, completely out of place. I knew who they were. People I'd met briefly on a holiday long ago in a completely different part of the country, who lived over 500 miles away. I had no contact with them. I didn't follow them online. I hadn't thought about them in years. But the image was precise. Three days later, I passed the café again — and there they were, exactly as I had visualised them. That moment was another affirmation of *precognition*, this time working in tandem with *clairvoyance*.

And it doesn't stop with this couple. I've started to sense when I'm about to cross paths with someone, whether I want to or not. The energy shifts. I'll be delayed, rerouted, or held back for some reason, and I get a distinct feeling: I'm about to see someone I know. I don't always know who, but when that feeling lands, it's never wrong. Without fail, someone familiar appears directly in my path, always in an unexpected place or context.

There was a deeper order. Something at work beneath the randomness. And the moment I admitted that to myself, everything else began to shift.

The Truth in the Mirror

Chapter 2: From Strategy to Subtlety

First steps into structured intuitive training.

The Annette Sessions

It was Annette Norbury who really opened the next door.

At the time, I didn't know her. Facebook recommended her to me as another animal communicator, even though I had never searched for animal communicators on Facebook. Annette is based here in the UK. I'd already had the reading from Julie in Utah, which had led directly to Askar's diagnosis and surgery. But I was still holding it all at a bit of a distance — stunned, curious, wary. I wasn't trying to "become intuitive." I just wanted to help my dog.

Annette's session came just after Askar's operation. He was recovering well physically, but I sensed that he was still holding something in his system. I couldn't explain how I knew that — it was just a knowing. And I suppose that was the beginning. The place where instinct started to become information.

I booked a session with her. And just like with Julie, the level of detail was astonishing. Annette was able to ask an animal to transfer their pain into her body so she could feel how badly they were hurting and where. This alone was extraordinary, and Annette is completely congruent so there is no mistaking that she has nothing to gain by making this up. What she shared with me from Askar wasn't just about physical symptoms — it was about emotion. Grief. Guilt. Loyalty. Things I hadn't expected to come through, but which made complete sense the moment she said them.

She described how Askar had been "holding on" during his pain, trying not to show it, trying to take care of me. And she picked up on something else too. She said he was "relieved" that I had listened. That I'd trusted what didn't make logical sense. That I'd acted.

The Truth in the Mirror

It felt like something turned over inside me in that moment. A click. A shift. I didn't have to believe everything. I didn't have to understand it all. But I couldn't deny that this had happened, that this communication, this exchange of energy, was *real*. And not only was the telepathy real, but the level of intelligence our pets hold is real, undeniably greater than our arrogant egos have ever given them credit for.

It was Annette who told me, quite directly: "*You should do this. You have the ability; we all do. It's just that most people don't have an open enough mind. You just haven't trained it yet.*"

And that was it. Not a lightbulb. More like a steady current that began to build in the background. It would still take a few more signs, a few more nudges. But that session with Annette planted a seed that would change the course of my life.

The Ziggy and Meg Case: A Turning Point in Belief

Annette showed me the way. Not just to the possibility of animal to human telepathy as a teachable skill, but to the truth that connection could cross time, space, and even death. Still, I needed more convincing.

Annette suggested, gently but with the clarity of someone who *knew*, that I try to communicate with her own dog, Ziggy, who had passed away a couple of years earlier. The idea that I could connect with an animal in spirit was, at the time, still beyond what I could truly accept. I had been raised in an atheist, academic household. There was no room in my logical mind for anything unseen or unproven. Even though my own lived experiences were beginning to challenge that worldview, the thought of reaching across the veil to a deceased animal felt like a stretch too far.

But curiosity has always been a more powerful force in me than skepticism. So I said yes.

The Truth in the Mirror

Although Annette had sent me a photo of Ziggy, I didn't have her photo with me the day I received my first communication. I didn't sit in meditation or light a candle. I simply went out walking. And then it happened.

A vivid image entered my awareness, unprompted, unforced. I saw a small, scruffy dog with a wiry black-brown coat. It was such a specific picture, like a memory I didn't know I had. With the image came a single name: "Meg."

I didn't think much of it at first. I had no reference point. I told Annette what I'd seen and what I'd heard and left it there. A few days later, she contacted me, sounding both intrigued and deeply moved.

She had come across a photo of a rescue dog online, a dog that looked almost exactly like the one I'd described. She and her husband went to meet the dog in person. And when they arrived, they were greeted by the woman fostering her.

Her name was Meg.

I remember the stillness in my body when she told me that. Not shock, not even disbelief, just a surprised settling. A knowing. A kind of *click* inside my system that felt as undeniable as gravity. This was real. This was happening. I hadn't made it up. The image, the name, the connection — I had tuned into something beyond the physical.

That one moment, so quiet, so humble, was the crack in the dam. It was the validation I needed to begin the journey in earnest. Annette had believed in me before I could believe in myself. And Ziggy, that gentle soul on the other side, had stepped forward to open the door.

I signed up for a 10-week course in animal communication not long after. The rest, as they say, is a different kind of history.

The Truth in the Mirror

Bridging Worlds: From Logic-Driven Career to Energy Awareness

If you had told me a few years ago that I'd be sitting here writing about spirit dogs, intuitive messages, and energy fields, I would've laughed you out of the room. I came from a world driven by logic, strategy, and results. I ran businesses across the globe. I understood shareholder value, turnaround plans, operational restructuring, and what it takes to stabilise a sinking ship. I didn't meditate. I didn't "feel into the energy." I certainly didn't believe in fairies.

But that's only half the truth.

Even at the height of my logic-driven career, I felt things. The tone of a meeting. The culture of a team. Whether a leader was genuinely aligned or masking something. I could walk into a boardroom and instantly sense whether a business was thriving or tolerating itself. I thought everyone could do that.

Turns out, they couldn't.

I just never had language for what I was sensing. It was intuition. Energetic intelligence. Subtle signals. Back then, I called it "instinct." But instinct is just intuition wearing a power suit.

I began to burn out, not because I wasn't good at what I did, but because I was only living from the neck up. My body was screaming. My emotions were trapped under decades of pressure to perform, to prove, to stay respectable. I had money, status, control, but I had lost touch with what made me *alive*.

The turning point wasn't dramatic. There was no collapse. There was just a quiet decision: *"I can't keep doing life like this."*

So I stopped. I took time out. And into that space came the whispers, gentle at first, then louder. Messages in dreams. Encounters with animals. Repeated signs. Synchronicities that logic couldn't explain.

The Truth in the Mirror

The deeper I went into energy awareness, the more I realised how *scientific* this all is — not in the clinical, peer-reviewed sense, but in the way that nature itself is scientific. Coherence. Resonance. Vibration. Frequency. These aren't "woo-woo" — they are laws of physics, expressed through biology.

And the more I studied, the more I saw that my work in business was never separate. It had *always* been energetic. Strategy is about alignment. Leadership is about emotional resonance. Culture is just the collective frequency of a company's people. I'd been reading energy my whole career, I just didn't know I was doing it.

Now I do.

So this work, the psychic sensing, the animal communication, the healing, the tarot, the synchronicities, it's not a rejection of my past. It's the completion of it.

I didn't leave my old world.

I bridged it.

The 10-Week Animal Communication Course and the Silva Method

Once Ziggy had shown up — and Meg had confirmed it — I knew I couldn't put this back in a box. I'd crossed a threshold. The part of me that had been trained to question everything, to filter life through reason and rationality, was now faced with something it couldn't disprove. It hadn't just been a lucky guess. It had been too specific. Too real. And too undeniable to ignore.

So I signed up for a 10-week animal communication course.

It was run online by a business in the U.S. I was in Scotland, but it didn't matter. Every week, we showed up via Zoom, me and a dozen or so others, and we practiced. At first, I was cautious. I'd get vague images: a ball, an animal drinking water. It was something. But not enough to satisfy my inner analyst.

The Truth in the Mirror

But each week, things shifted. The detail deepened. The information started coming through with texture. A hand under a horse's belly. The memory of a sore tooth. An alleyway full of raccoons. At first, it felt random. Then it didn't.

One of the biggest surprises wasn't just what I was perceiving, it was that other people were perceiving it too. We were all reading the same animal each week, and even though we used different words or symbols, the energetic fingerprint matched. The owner would validate our messages, and that validation fed my courage.

We weren't just imagining things.

We were receiving.

Around the same time, I enrolled in something called the Silva Method, a structured training in intuitive perception and energy awareness. It was a different flavour of the same truth. This time we were working with human energy fields, tuning into people on the other side of the world, sensing medical conditions, emotional states, and relationship dynamics. Again, the messages came. Again, they were verified. I began to trust that what I was receiving wasn't wishful thinking. It was energetic information. And it was verifiably accurate.

I remember one practice partner asked me to tune in to her brother-in-law in the U.S. I saw him clearly: diabetic, stocky build, a certain way he carried his weight. She confirmed it all. Another time, I saw her friend roller skating and mentioned a spinal issue. "She's obsessed with skating," she told me. "And yes — she's got a bad back."

The most moving reading, though, was with a dog whom I couldn't quite see... I kept seeing an Alsatian, but it wasn't the dog I was reading for. I then got a flash of light — sudden, bright — and the knowing came: the dog I was reading was blind. The Alsatian was her companion, her guide. That one floored me. Again, it was all confirmed by my training partner.

The Truth in the Mirror

With every session, something rewired inside me.

This wasn't just about animal communication anymore. It was about how energy moves. It was about how we *know* things, before we know how we know them. It was about the truth that the body, the mind, and the field around us are always talking to one another. Always listening. Always transmitting.

I hadn't abandoned logic. I'd just found a deeper layer beneath it.

Shared Perception in Group Training

If there's one thing that dismantles the myth of "*you're just making it up*," it's shared perception.

When you're sitting in a Zoom room with at least twelve other people, all tuning into the same animal you've never met, and all of the group gets the same message in different words... something shifts in you. You stop doubting. You start recognising.

In those early training sessions, the animal's energy would land differently in each of us. Someone might feel a tightness in their chest. Someone else might get a visual of a darkened barn. Another might hear the word "dust" or "old hay." But when the owner came on camera to confirm yes, she'd just moved him to a dark stable because the old barn had been too dusty, it was like our intuitive threads braided together into one undeniable rope of truth.

It wasn't consensus. It wasn't guesswork. It was resonance.

This was perhaps the most startling part of learning in community. Not just that my own senses were working, but that a shared energetic field exists. And we can all access it.

There was a week where we connected with a dog, and three of us independently described a blue ball, same shade, same squeak, same obsession. Another time, I described a hand reaching under a horse's belly. One of the other women said, "I

The Truth in the Mirror

saw that too." The owner confirmed: that was a special trust-based routine she did every morning, and only with that particular horse.

It was humbling.

The group energy didn't just validate the process, it amplified it. When we tuned in together, it felt like my clarity sharpened. It was as though we opened a shared frequency band, and the more open-hearted we were, the stronger the signal got.

Sometimes, we'd finish a session and just sit there in silence, blinking at each other through the screen. Not because we were confused, but because we knew something sacred had just happened. And it had happened through all of us.

This wasn't the world I was trained in. This wasn't linear. It wasn't logical. And yet, it was precise. Verifiable. As accurate at times as any spreadsheet or MRI scan could offer.

It was a different kind of intelligence. One that lives in the space between us.

And once you've felt it, you never forget.

Weekend Intensives and Early Training with Erin Furman

If the group course opened the door, the weekend intensives and out-of-class training sessions blew it off its hinges.

By the time I met Erin Furman on the animal communication course, I already knew animal telepathy was real thanks to Annette. But I didn't yet know what it could become. Erin was the first person I trained with who met me in the full depth of it — no hedging, no holding back. Just truth. Directly from the animal.

We began practising together outside of formal training, swapping sessions and trusting the process to lead. And it did — again and again. It led us into territory neither of us could have imagined.

The Truth in the Mirror

We'd sit quietly in our respective homes, across oceans, yet just down the Zoom line, and tune into the same animal. Within moments, we'd both start describing identical sensations, behaviours, images, as well as receiving some specifics that were individual to just ourselves, yet equally verifiable by the animal's owner.

That was the level of specificity we were reaching. And it wasn't random. It came from disciplined attention, honest feedback, and deep energetic trust.

The additional weekend courses we attended, long, immersive containers of structured telepathic practice, were foundational. These weren't airy-fairy escapes from reality. They were full-bodied plunges into it. Raw. Often emotional. But always, always growing our skills.

What I learned with Erin was that animal communication isn't about being 'gifted.' It's about becoming trustworthy.

Animals don't care about your CV. They care about how still you are. How clean your intention is. Whether you will hold what they give you with respect, or twist it to suit your own story.

Training with Erin was a masterclass in this level of integrity.

Together, we explored subtleties that most people miss: the difference between emotion and projection; the line between empathy and invasion; the need to stay grounded in your own body while reading another's. We learned the signals that meant we'd veered off-course. We learned how to come back.

We learned, most of all, how to listen. And that kind of listening, the kind that changes your nervous system, isn't something you can unlearn.

It re-patterns you from the inside out.

The Truth in the Mirror

Chapter 3: The Evidence Grows

When the data becomes undeniable.

Remote Viewing Through Animals' Eyes — The Cat and the Wagon Wheel

There's a moment in every intuitive journey when the mind runs out of plausible denials. When "maybe I imagined it" no longer holds. When the detail is too sharp, too specific, too accurate to ignore.

For me, that moment came when I became the eyes of a missing cat.

At the time, I was practising weekly with Erin and another woman from my animal communication course. She lived on a large rural property, acres of land, scattered outbuildings, plenty of space for a wandering feline. But her cat had vanished. No trace. No clue where she might have gone.

She reached out, asking if we could help. I had no experience tracking missing animals, but I treated it as another opportunity for training. I didn't know the layout of the land. I hadn't seen photos or maps. It was just me, a Zoom screen, and a simple invitation: "*Can you tune in and see anything?*"

I closed my eyes and dropped in. What happened next felt like slipping through a membrane.

Suddenly, I was inside the cat's perspective, low to the ground, confident, weaving through obstacles. Then I saw it. Clear as day: a large wooden wagon wheel, old-fashioned, propped up beside a tree. Next to it, an old tractor, dusty, unused, a relic. These weren't metaphors. They were real, physical objects. I was seeing them through the cat's eyes.

I described what I saw in real time.

The owner paused. "Hang on," she said. "I'm getting in the car."

The Truth in the Mirror

Ten minutes later, I received a photo. It was her land, a different part of the property from her house, and to reiterate, a place I'd never seen. And there it was: the wagon wheel. The tractor. Exactly as I'd described.

That moment cracked something wide open.

Because this wasn't just intuition — it was evidence. Live, timestamped, verifiable remote viewing. The cat's sensory data had travelled across the Atlantic and landed in my awareness — or perhaps more accurately, I had somehow landed inside hers.

And if that's possible with one animal... what else is?

That question became a portal. One I've been walking through ever since.

Reese & Eric — Cherries, the Truck, and Country Music

There are stories that make you smile, and there are stories that stop you in your tracks.
This was one that encompassed both.

It happened during a practice session with Erin. She asked if I'd like to connect with her dog Reese, who had passed a few years earlier. I'd had in front of me a photograph of Reese, something we used to help anchor the animal's energy, but I knew nothing about her life, her personality, or her relationship with Erin's husband, Eric. I had said hello and waved to Eric once or twice on Zoom, but I knew very little about him. I was still over in Scotland, Erin and Eric in central USA, and we had never met in person.

And yet, within moments of tuning in, a sequence of images began to arrive. Vivid. Precise. Full of feeling.

The first one took me completely by surprise: Eric sitting by a lake or river, fishing, and Reese, joyful and excitable, bounding through the water, trying to reach the

The Truth in the Mirror

fish before they landed. I shared the vision aloud. Erin looked to Eric, who was sitting beside her, just off-camera.

He confirmed it instantly.

That's exactly what they used to do, he said. Eric and Reese. Fishing trips together, just like that. Reese would get so excited, racing into the water as the fish came in, always trying to "help." It was one of their happiest shared routines.

The next image followed almost immediately, just as strong. A flatbed truck with a wide, single bench seat across the front. Reese was perched on the seat, pressed up against Eric, and he had his arm around her. I heard him singing too, singing at the top of his lungs, country and western music, full-hearted and unfiltered.

Again, Eric nodded. That was their truck. That was their seat. That was exactly how they used to drive around together, Reese curled into his side, him belting out country tunes, no care for who heard him. The kind of ordinary magic you never forget.

Then came the third and final moment.

Eric asked: "Does she have any signs for me?"

The image that came through wasn't just symbolic, it was specific, almost tactile: bright red cherries, fresh and juicy, with the stones being removed. I could feel the ritual in it. The tenderness. Like it was a quiet act of love repeated many times.

Eric's eyes widened. He had completely forgotten, until that moment, that he used to do exactly that. He'd buy paper bags of cherries, remove the pips one by one, and feed them to Reese. She loved them. It was their thing.

There was no way I could have known that. No way to guess. The sequence, fishing, the truck, the music, the cherries, was too specific, too textured, too intimate to be anything but real.

The Truth in the Mirror

This was Reese. Still connected. Still loving. Still present. Even though she was no longer in her physical body.

That session changed me. Not just in my growing trust in what I was receiving, but in something deeper: my understanding of presence. Death, I realised, does not erase love. It simply alters the way it speaks.

This session also deepened my curiosity. This was real and tangible; there was no question that I was picking up on information from somewhere. But where? Was it the energy and messages from a dog in spirit, or were they dormant memories from within Eric, ones he would never have recalled without prompting, but which were accessible nonetheless through the telepathic transfer of energy?

Ben — The Horse with the Hidden Illness

Some readings come through like a whisper. Others arrive with the weight of knowing, visceral, undeniable, and not always easy to carry.

Ben was one of those.

During a quiet period between formal training sessions, Erin asked if I wanted to do a practice reading for a horse. She didn't tell me anything, only that the vet had been called, and that it was a chance to see if I could pick anything up.

I wasn't trying to diagnose anything. That's never the role of a communicator. But I tuned in, as I'd been learning to do, and let the sensations unfold in my own body.

What I felt was immediate and distinct, although slightly challenging to describe in words: a deep pressure, almost like a cavernous tension at the back of the throat and up behind the palate. A sense of emptiness somewhere behind the nose, higher than the mouth, but not in the lungs. There was also a feeling of loss, of something draining away. Not blood, exactly, but condition. Strength. Vitality.

And then came the image, of Ben standing slightly off balance, not outwardly distressed, but dulled. Weary. His energy felt frayed.

The Truth in the Mirror

I scribbled everything down. The location of the sensation. The odd quality of the depletion. I noted the sense of weight loss and the distinct feeling that whatever this was, it wasn't muscular or skeletal, it was something internal and hard to reach that was making him lose condition.

After I recounted my findings, Erin confirmed that Ben had been having intermittent nosebleeds. He was indeed losing weight. The vets were puzzled.

Not long after, a devastating diagnosis came back: a rare cancer, affecting the blood vessels in a way that matched what I'd felt. It was causing the bleeding. The energy drain. The decline.

Ben was sadly put to sleep shortly afterward.

This was the first time I'd sensed something so serious before it had been confirmed by medicine. And while it wasn't a joyful validation, it was a real one. I hadn't known anything about the case, but my body had received the message clearly and accurately. Not to replace the vet, but to add a layer of insight that no scan or test had yet revealed.

What that session taught me is that animals do tell us when something is wrong. Not always in ways we expect, but through sensation, image, and transmission. And when we slow down and learn to listen with our whole body, that language becomes increasingly accessible.

For me, Ben was a turning point, not just in terms of what I could perceive, but in how I understood the sacred responsibility of this work.

It's not always light and easy. But it's always real.

Freya — The Missing Cat Under the Decking

Some animals shout. Others whisper. And sometimes, the message is a still image so crisp and quiet that it holds its own kind of urgency.

The Truth in the Mirror

Freya is a house cat. Sixteen years old, sensitive, and deeply bonded to her human. She had rarely been outside on her own, then one day she disappeared.

Her disappearance sent a ripple of worry through the family, and Erin contacted me. She explained the situation briefly and sent a photograph of Freya. That was all I had. But as has so often been the case, the image was enough.

The moment I saw Freya's face, I received a clear message. The word came first — "measles" — which seemed odd. I looked it up and found that one of the feline viruses associated with that term was feline panleukopenia, a type of parvovirus. Cats affected by this virus often seek solitude. They hide.

Then came the image.

I saw Freya from behind, nestled under something wooden. Lying on her side, her back was turned to me. She was lying on dark earth beneath slats, perhaps a deck or a porch. Light was filtering through a narrow gap at one side and there were pillars holding up the wooden structure. The air felt cool, but not dangerous. She was alive. Quiet. Tucked away.

I described everything to Erin and asked her to relay it to her sister, Freya's human.

Several hours later, confirmation came through: a neighbour, five doors down, had spotted Freya. She was exactly where I'd seen her, under decking, in the shadows, hard to spot because her back had been turned. Erin's sister retrieved her just in time.

The story didn't end there.

Even before I knew she'd been found, I'd received a second image: Freya at the vets. Lights, instruments, the clinical feel of a table. I didn't say anything at the time because it hadn't been confirmed, and I didn't want to give false hope. But when I later learned that Freya had indeed been taken straight to the vet for fluids and care, everything clicked into place.

The Truth in the Mirror

This was one of those moments where the precision of the image left no doubt.

Freya had communicated her location, and her state, to me visually and verbally with quiet clarity. All I had to do was receive it.

These moments don't just affirm that telepathic communication works. They remind me that what we call "lost" is sometimes just temporarily out of reach, waiting for someone to listen in the right way.

Feeling Pain in a Dog's Ear – Saving Jasmine's Life

Some animals speak in symbols. Some send images. Some push through in silence so loud it hurts.

Jasmine was one of those.

A sweet, sensitive little dog who had recently started urinating indoors, something completely out of character for her. There were no signs of distress, no indications of illness, just behaviour that didn't fit.

Erin and I tuned in to her, separately, using only a photograph. This is one of the most powerful ways to train, reading without outside information, then comparing notes. When impressions match, it builds trust. Not just in each other, but in ourselves.

And that's exactly what happened.

Both Erin and I picked up inflammation, specifically in Jasmine's lower back, and both of us received the clear sense that she was dealing with a urinary tract infection. That alone would have been helpful, but Jasmine wasn't finished.

There was something else.

While tuning into Jasmine's energy, I experienced a sharp, unmistakable pain in my right ear. Not metaphorical. Physical. It was a kind of pain I couldn't ignore, and it didn't belong to me. Then, more quietly but unmistakably, I heard the words: "Help me." Not out loud. Not even fully formed. But as a psychic impression of a silent scream, so urgent and so raw that it cut straight through the noise.

Jasmine's humans were already heading to the vets. They asked for the ear to be examined, just in case.

What the vet found shocked them, and me.

Jasmine did indeed have a urinary infection. But they also discovered one of the worst inner ear infections they'd ever seen in her right ear. Completely invisible from the outside, and likely incredibly painful. It was treated immediately with antibiotics. Jasmine made a full recovery.

She hadn't been able to show it in the ways we expect, but she had asked for help. And she was heard.

That's the quiet miracle of this work. The cross-correlation between two independent intuitives isn't just about validation — it's about building confidence, deepening trust, and letting the animals guide us with clarity and care.

Harriet: A Telepathic Connection That Changed Everything

A friend reached out to ask if I would do an animal communication session for their friend's horse. I don't publicly offer animal communication sessions, though I'm trained in it, I pursue it out of curiosity, not as a profession. There are many professionals who dedicate years to honing their skills, and I never want to detract from their work. I normally recommend Erin and Annette if someone enquires. However, I also know that I'm naturally gifted in this area, and I now teach a foundation-level course on how to learn this skill in collaboration with Erin.

As soon as my friend mentioned her friend's horse, I started receiving information. It was immediate, unprompted, and detailed. By the time I officially agreed to the

The Truth in the Mirror

session and received a picture of the horse, I had already written eight pages of notes. Before the owner, Selina, even sent her list of questions, I had already answered them.

The horse, Harriet, was clearly communicating. But here's the question: Did Harriet already know what Selina wanted to ask? Or was I picking up on Selina's thoughts before she had even verbalised them? This is one of the aspects of telepathy that fascinates me, the interplay between animal consciousness, human intention, and energy transmission.

To be clear, I have never met Selina or Harriet. They live 150 miles away from me. I'd never seen any of Selina's social media profiles. I'd never asked, and nor did I want to. What would be the point of faking something so vital, so meaningful, and so full of scientific implication?

Before I even received Selina's questions, I had documented pages of notes based on what Harriet was communicating to me. When Selina eventually sent her list of questions, I realised Harriet had already provided the answers.

Harriet shared specific information, issues with her saddle, an old accident that left a scar, and pain in a specific leg. Selina had said there was "a leg issue" but hadn't told me which leg. I pinpointed it, along with tension in Harriet's neck, which Selina could also verify. All of it had come through before any of that detail was shared.

I shared my notes with Selina. And to support this even further, I have a photo of the handwritten notes I took before receiving Selina's questions. This matters. Because it shows the information was coming directly from Harriet, not from a process of deduction, guessing, or subtle prompting.

The day after our session, an image kept returning to my mind. Repeatedly, I saw Harriet standing on a grassy hill, staring directly into the camera, with heavy grey clouds rolling behind her. It was vivid. Persistent. Undeniable.

The Truth in the Mirror

I didn't know what it meant, but I described it to Selina via a voice message.

Her reply left me speechless.

She sent me a photograph she'd saved to her phone two years earlier, an image of a Clydesdale horse like Harriet, standing on a grassy knoll beneath grey clouds. It had been on her phone for two years before rescuing Harriet. This was her vision board, her manifestation image. She knew she wanted a horse like that.

In the original image, the horse was looking away. But Selina then sent a video that captured the first moment Harriet, after being rescued, ever truly engaged with her. Her daughter was brushing Harriet when Harriet turned, locked eyes with Selina, and held that gaze. It was the moment of connection. The moment of acceptance.

The image I had seen wasn't just symbolic — it had already happened. Harriet had shared it with me.

Weeks later, I had another spontaneous experience. I had a strong sense that Selina would be in touch. Within hours, she messaged, Harriet was distressed, feisty, and unsettled.

As soon as I tuned in, I saw an image of Harriet's hooves, almost warping in shape. I felt energy travelling up into her legs, like a vibration coming from the ground. I said, "There's something coming up from below, some kind of energy disturbance."

We talked it through. There was an RAF base nearby. Were there aircraft drills? Or perhaps forestry machinery? And then we figured it out. The ground was saturated from days of rain, and a neighbouring field had a makeshift electric fence. The wet ground was conducting the electrical current, sending tiny shocks into Harriet's hooves.

The Truth in the Mirror

Selina moved Harriet to a different field. The change was immediate. Harriet calmed. Her agitation vanished. Selina later sent me a photo of the temporary electric fence that had been shorting.

In the same session, I'd also seen another image, Harriet's front hooves. As if I were standing right in front of her, noticing small chunks missing from the inside edges. This image felt different. The direction of the vision, the hooves pointing towards me, suggested it may have come via Selina's consciousness, not Harriet's. I wasn't sure. I held off mentioning it.

But later that day, Selina messaged me unprompted: *"Her feet are in rough shape. I've been trying to get hold of the farrier."* Along with the message, Selina texted a photograph.

It was exactly what I'd seen. The image was true.

The first reading session I described with Harriet transformed the relationship between her and Selina. Until that point, Harriet had been difficult to handle, a giant horse with an even bigger personality. Bold, temperamental, and unafraid to throw her weight around, she often challenged Selina's efforts to connect. But Selina later shared that the day after our session, Harriet's energy completely shifted. She became calmer, softer, and more amenable, settling into a more relaxed state. Something changed for both of them that day, energetically, emotionally, and relationally. Since then, they've grown together in an entirely new way, learning to trust, listen, and enjoy each other more deeply.

The Science of Telepathic Communication

I've had many more experiences like these, not just with Harriet and the other case studies above, but with lots of other animals and their humans. The challenge is that telepathy is still marginalised. I grew up in a scientific, atheist household. I used to believe that if something couldn't be proven, it couldn't be real. But I now know intelligent energy exists — beneath and beyond our current understanding.

The Truth in the Mirror

Telepathy isn't supernatural. It's biological. Emotional. Energetic. It's not a gift. It's an ability, a faculty that lies dormant in most people due to suppression, ridicule, or cultural conditioning. I learned it because I stayed open and trained. I don't have a different set of body organs from you, at least not as far as I know! And if I can do this, you can too.

We already know the human body transmits and receives data. The gut, the so-called "second brain", processes vast sensory input and communicates with the brain via the vagus nerve. Over 90% of serotonin is produced in the gut, not the brain. Our bodies know things before our minds do. We are built for intuitive connection.

So why do we assume we've lost this capacity? We haven't. We've just forgotten how to listen.

The Problem with Scientific Dismissal

Skeptics love to say, "That's not possible." But that isn't science — that's bias. The scientific method is about inquiry, not arrogance. If something keeps happening, with repeatable patterns and predictive outcomes, that's data.

So far, we just haven't built the right machines to measure what we already know to be real. But to dismiss a phenomenon just because it doesn't yet fit into our framework? That's not rational. That's intellectual cowardice.

We've been trained to silence our inner voice. To distrust emotion. To dismiss the body's wisdom. But there is a cost to that. When we cut ourselves off from intuitive knowing, we also cut ourselves off from our wholeness, our clarity, and our natural protection against manipulation and disease.

This is not about belief. It's about observation. I don't "believe" in telepathy. I know it, because I live it.

Animals are communicating with us all the time. Through energy, emotion, sensation, and silence. The real question is: Are we willing to listen?

Part Two: What the Data Reveals

The Unseen Architecture of Human and Animal Communication

Chapter 4: How Animal Communication Works

The multisensory mechanics of intuitive connection.

Animals Send Thoughts, Words, Images, Feelings, Knowing — Even Tastes and Smells

If you've ever had a dog stare at you intensely and somehow *know* exactly what they're asking for, whether it's a walk, a treat, or just to come closer, you've already had a glimpse of animal communication. What most people don't realise is that this experience isn't just emotional or behavioural. It's energetic, multisensory, and often telepathic.

Animals communicate in rich and complex ways that go far beyond body language and vocalisation. Through the work I've done with horses, dogs, cats, and even animals in spirit, I've learned that they send thoughts, visuals, sensations, emotions, and sometimes even smells or tastes. And, perhaps most surprisingly of all, humans can receive them.

This isn't imagination. It's not anthropomorphism. It's not wishful thinking. It's transmission and reception. In some ways, it's no different from tuning a radio dial, only instead of FM or AM, we're working with frequencies of consciousness and coherence.

A Language Without Words

When an animal sends a thought, it doesn't always arrive as a sentence. Often, it comes as a wordless concept, a knowing, or a picture. You may hear the phrase "claircognisance" to describe this, but it's simply a flash of internal understanding. Like a memory you didn't create.

When an animal sends an image, it may show up as a still visual, or as a full motion video clip inside your mind. You don't "see" it with your eyes, but it has visual

The Truth in the Mirror

weight and detail, colour, perspective, action, and sometimes even symbolic meaning. These are the hallmarks of clairvoyance, another term often misunderstood as mystical. In practice, it is just visual reception without the use of your physical eyes.

When an animal sends an emotion, you may feel it land in your own body as if it were your own, grief, panic, affection, irritation, or fear. This is common. It's also why animal communicators must be exceptionally grounded and self-aware: so we don't confuse what's ours with what we're being shown. This emotional transmission is what's often called *clairempathy*, and it's not just sensitivity, it's shared feeling.

Smells and tastes are rarer, but when they come through, they're unmistakable. In one session I'll never forget, a vivid, metallic, almost blood-like taste flooded my mouth and nose while connecting with a cat who'd recently been away from home hunting. I had no way of knowing that, but the taste was undeniable. This is known as clairsalience (clear smelling) and clairgustance (clear tasting), and while less common, they're real, and verifiable.

Animals Speak in Frequency

What we're talking about here is not imagination. It's reception. Animals don't "speak" in the way humans do, but they transmit nonverbal information vibrationally, across multiple sensory pathways. And when humans learn to quiet their minds and open their systems, we can receive those signals clearly and consistently.

This type of communication isn't supernatural. It's natural. It's just not currently part of mainstream understanding, yet. Our biggest challenge beyond receiving the information is to then translate it into an accurate verbal representation, undistorted by our ego.

In the sections that follow, I'll explore how we receive these signals, through our body's own intuitive technologies: the clairs, yes, but also the vagus nerve, the electromagnetic field of the heart, the enteric nervous system in the gut, and the DNA coils that may function as resonance receivers.

But it starts with this:
Animals communicate telepathically and energetically all the time.
The question isn't whether they're speaking.
The question is whether we're listening.

The Eight Clairs in Animal Communication

How animals speak through your subtle senses

Most people think of intuition as just a gut feeling. But what if that feeling is only one doorway into a much deeper, richer language?

Intuitive animal connection happens through the body, not just the brain. It's multisensory, subtle, and deeply natural. What many call "psychic" is, in truth, a set of innate human abilities that have gone unused. We are, after all, animals more than we are a mental ego mind.

You were born with all of these abilities.
Some may speak louder than others.
Some may feel unfamiliar at first.
But they're there — waiting to be remembered.

Here are the Eight Clair Senses, as they show up when animals speak to you:

1. Clairsalience – *Clear Smelling*

Animals often transmit energy through scent, especially when communicating about trauma, memory, or warning. You might suddenly smell hay, biscuits, wet fur, aftershave, or even something unpleasant, such as an infection or blood. There is no physical source, just a message arriving through the nose, even when you are not physically in proximity to the animal.

The Truth in the Mirror

This sense often activates when an animal wants to remind you of a shared ritual or point to a problem in their environment.

2. Clairgustance – *Clear Tasting*

Rare, but powerful. You might suddenly taste cherries, kibble, meat, antiseptic, blood or bile. It often comes when animals are recalling specific treats, medication, or foods that were significant, or when there's an issue with digestion or toxicity.

Taste is deeply tied to memory. This is their way of transmitting emotional imprint through flavour.

3. Clairvoyance – *Clear Seeing*

Animals very often communicate in pictures, not words. You might see a blue collar, a red ball, a muddy field, a particular tree, or even an injury they once sustained. Sometimes it's symbolic, like a cage or a door, and other times it's literal.

Clairvoyance allows you to "see through their eyes," witness their world, travel with them, live in real-time through remote viewing, or glimpse their past.

4. Clairaudience – *Clear Hearing*

You may hear a name, a random word, the sound of a leash, a bark, a whisper, or a full sentence. It might sound like your voice, or theirs. Some animals send song lyrics, human phrases, or the tone of voice their person uses.

Clairaudience is often how animals let you hear something that was said to them, or how they send back something they've been holding onto.

5. Clairsentience – *Clear Gut Feeling*

This is the most instinctive of all the clairs, a clear, bodily hit of truth. It's the "yes" or "no" feeling in your gut. The sense that something is off. In animal communication, it often shows up as a general sense of discomfort, restlessness, or calm that isn't yours.

The Truth in the Mirror

Clairsentience is not emotional, it's energetic. You feel the truth of something through your gut or skin, without any accompanying story.

6. Claircognizance – *Clear Knowing*

This one drops in like a fact. You suddenly "just know" that the animal is grieving, that something happened during a thunderstorm, or that their bowl was moved, and they're annoyed about it.

There's no logical trail, just certainty. It's a knowing that arrives fully formed. This is often a dominant channel in people with sharp strategic minds or fast pattern recognition.

7. Clairempathy – *Clear Emotional Feeling*

Here's where the heart opens. Clairempathy allows you to feel the animal's actual emotions and physical ailments in your own body, their loneliness, joy, frustration, love, grief. This is how an animal can "send" you sadness or fear, pain or inflamation, and you feel it as though it were your own.

It's not a thought, or even a gut feeling, it's full emotional resonance.

8. Clairtangency – *Clear Touch*

This sense allows you to feel textures, heat, cold, or pressure from an animal, even remotely. You might sense a tightness in your jaw, a dull ache in your hip, or warmth in your hands as if they're pressing into fur.

Some communicators use this to scan the animal's body and pick up areas of pain or blockage. It's especially common in energy workers and healers.

All of these clairs are natural. They are not mystical or elite. They are your body's extended sensory range, refined through quietness, trust, and repetition.

In animal communication, they allow you to receive thoughts, memories, and signals from the animals who are always speaking.

The Truth in the Mirror

Not in English.
Not in sound.
But in energy.

DNA Coils, Vagus Nerve, Gut and Heart Fields as Sensory Hubs

Your body was built for subtle perception.

Intuitive communication isn't just a skill.
It's a function, deeply biological, intelligently designed.

We often imagine psychic or energetic sensing as "otherworldly," something floating outside the body. But the truth is, it happens *within and through* the body, through your physical structures, your electromagnetic field, your cellular memory.

When animals, including other humans, speak to us, we don't just "think" the information.
We feel it.
We know it.
We resonate with it.
And that resonance is made possible by very real, very measurable parts of our physical form.

Let's explore a few of them.

DNA Coils: The Living Spirals of Memory and Frequency

DNA isn't just a static genetic code.
I believe it's a spiral antenna.

In what I call the Heather Ogilvie Theory of Life Force Transmission, I propose that energy moves through the body through resonance across *spiral biological structures*, most notably DNA, the gut lining, and the brain's neuronal coils.

The Truth in the Mirror

Double helices naturally generate and receive electromagnetic signals. In physics, spiral coils are known to amplify and conduct energy, this is the basic mechanism of radio transmitters, transformers, and even wireless communication. It's not a leap to see your DNA as a biological coil that picks up subtle information, including the emotional or energetic signals of others.

Rather than a linear flow, this theory suggests a multidirectional, dynamic resonance, energy "jumping" across coils like a transformer stepping up or down voltage. These living spirals may act as conductors of intuition, of soul connection, of interspecies communication.

This isn't supernatural. It's an emerging view that integrates lived experience with the physics of resonance, and begins to explain how "knowing" arrives before logic does.

The Vagus Nerve: Pathway of Emotional and Energetic Data

This is your body's major information highway, connecting brain, gut, heart, lungs, and more. It's the vagus nerve that gives you a lump in your throat when you're about to cry or makes your stomach flip when you walk into a room that doesn't feel safe.

In animal communication, it's often the vagus nerve that carries the first signal:
A subtle tightening in the chest.
A flip in the gut.
A pressure or pain in the skeletal system.

This is not metaphor. It's biology. The vagus nerve processes input far faster than the conscious mind. Animals use this system constantly. So do we — we've just been trained to ignore it.

The Gut: Sensory Powerhouse and Second Brain

The gut isn't just for digestion. It's a neural network.

The Truth in the Mirror

Known as the "second brain," the gut has over 100 million neurons, more than the spinal cord, and is capable of processing complex data, storing memory, and sensing emotional truth.

More than 90% of serotonin is made here.
Signals travel from the gut to the brain first, not the other way around.

So when you feel an animal's truth land in your stomach, a yes, a no, a sadness, a plea, that is your gut responding to a real frequency. Not imagined. Not metaphorical. *Real*.

The Heart Field: Electromagnetic Intelligence

The heart's electromagnetic field is 60 times stronger than the brain's. It radiates 360 degrees around your body, like a broadcast signal.

When two beings, whether that's human and animal, human and human, or animal and animal, come into proximity, their fields interact. This is why you can feel trust, tension, grief, or love before a single word is spoken. Your heart is reading the other's field.

In intuitive communication, the heart often receives emotion before the mind can translate it. This is why you may start crying during a session, or feel a sudden rush of warmth, or be overcome with stillness.

You are receiving the animal's truth through resonance, heart to heart.

You don't need to believe in anything mystical to understand this. You simply need to remember that the body is intelligent, far more than we're taught.

Animal communication doesn't bypass the body.
It begins there.

The Truth in the Mirror

And when you trust the messages coming through your DNA, your gut, your vagus nerve, your heart, you begin to realise this is not "woo-woo". This is how we're wired.

Intuition as Electromagnetic and Resonant, Not Imagined

We don't imagine intuition.
We receive it.

Despite centuries of dismissal, intuition is not a vague feeling or a lucky guess. It's a reception. A broadcast. A frequency match.
And it operates through one of the most fundamental forces of nature: electromagnetism.

Electromagnetic resonance governs everything from wireless communication to the behaviour of atoms. Tuning forks, radios, even musical instruments all rely on resonance, one frequency vibrating in harmony with another.

So does your body.

When your energy field comes into contact with another, a human, an animal, even a place, your system scans for coherence.
It listens. It feels. It reads.

And when there's resonance, a match, a signal — that's what we call intuition.

It might arrive as a thought.
It might feel like a physical tug in your stomach.
It might come as a flash of insight or a flood of emotion.
But the transmission is real.

You don't have to try to make it happen.
You tune to it, by stilling the noise, softening your logic, and allowing your system to notice what it's already perceiving.

The Truth in the Mirror

This is why animals seem so "in tune", they are not overriding their signals with ego, analysis, or societal conditioning. They receive, respond, and relate directly from the field of resonance.

And when we return to this way of listening, we remember something ancient and essential:

That our intuition isn't a fluke.
It's a frequency.
And we are built to respond.

Aligns with Mirror Neurons, Quantum Coherence, and Brainwave Syncing

You don't need to be "psychic" to understand animal communication.
You only need to be human.
Because the very science of being human already supports how it works.

Let's break it down.

Mirror Neurons: The Science of Empathic Connection

When you see someone yawn and feel the urge to yawn too, that's mirror neurons in action. These are specialised cells in the brain that activate when we observe the experience of another. They don't wait for logic or language. They *feel* energy. This means we're wired to resonate — not just with people, but also with animals.

When an animal feels something deeply, and we are still enough, our mirror neurons can fire as if we were feeling it ourselves.

That moment when a horse's grief brings you to tears, or a dog's joy fills your chest — it isn't imagined.
It's your nervous system joining theirs in real time.

The Truth in the Mirror

Quantum Coherence: Shared Fields of Possibility

Quantum physics teaches that particles can become "entangled", influencing each other's states across vast distances.
It also shows that observers are not separate from what they observe.
In the quantum field, everything is connected.

When we tune into an animal remotely, with no physical contact and often no prior knowledge, and receive accurate, specific information, we are likely accessing a form of quantum coherence. Two beings become temporarily entangled in a shared field of attention.

Not through force.
Through trust.

Brainwave Syncing: The Power of Presence

When two people spend time together, especially in a calm, connected state, their brainwaves can synchronise. This phenomenon has been observed between mothers and infants, as well as among musicians, therapists, and their clients.

It's no surprise that animal communicators report falling into altered states of consciousness when connecting with animals.
Meditative. Trance-like. Expansive.
The brainwaves slow down, into alpha or theta, and in that space, intuitive data becomes clearer.

Animal telepathy often happens in this zone of synchrony.
The brain aligns.
The energy stabilises.
And the truth, images, words, feelings, knowing, comes through.

This isn't supernatural.
It's natural.
We've just forgotten how to listen to the instruments inside us.

The Truth in the Mirror

Grounding "Woo-Woo" in What's Already Observable in Science

There's a tired, arrogant, outdated assumption that intuition, energy, telepathy and animal communication are "woo-woo", vague, unprovable, fringe.

But here's the truth:
Most of what people label as "woo-woo" is simply wisdom we haven't learned to measure yet.

We didn't always have machines to detect brainwaves, or microscopes to view bacteria.
We didn't always know the Earth wasn't flat, or that invisible germs could cause disease.
Science evolves. And it always starts with observation.

What I'm sharing in this book — what I've lived and verified over and over — doesn't sit in opposition to science.
It sits ahead of it.

We can already observe:

- The electromagnetic fields of the heart and brain
- The structure and possible function of DNA coils as energetic receivers
- The role of the vagus nerve in processing emotional and somatic signals
- The presence of mirror neurons and their role in empathy
- The ability of the gut to process truth before the mind
- The measurable shifts in brainwaves during altered states

None of this is made up.
None of it requires belief.
It only requires curiosity.

The Truth in the Mirror

Animal communication, intuitive knowing, and subtle energy sensing are not mythical.

They are biological.

And while the language might sound mystical to some, the *function* is very real.

We're only scratching the surface of what human and animal intelligence can do, especially when we stop pretending that logic and human intellect are the only form of knowing.

You don't have to stop being intellectual to become intuitive.

Because the greatest intelligence includes the intuitive.

It listens to data *and* instinct.

It honours patterns, fields, energy, emotion, and truth — even if they arrive wordlessly.

It's not "woo-woo".

It's woven.

Right into the fabric of who and what we are.

The Truth in the Mirror

Chapter 5: Beyond Belief – Verifying the Unseen

Tightening the lens: data, discernment, and death.

Cross-Validation from Multiple Readers at Once

One of the most powerful tools we have in intuitive training is cross-validation.

It's what takes a personal, unprovable experience and places it squarely in the realm of replicable data. When two people, or three, or more, tune into the same animal or person, and independently receive overlapping information, you're no longer working with vague guesswork. You're working with something deeper. Something consistent. Something real.

I've seen this again and again.
In my own practice. In teaching others. In training myself to trust the unseen.

The example I've already shared of Jasmine stands out.

Jasmine was a gentle dog whose behaviour had suddenly changed. She'd started urinating inside the house, something completely out of character. There were no obvious signs of illness, and nothing in her environment had shifted. Her humans were concerned, and so Erin and I both tuned in, separately, to see what we could find.

We didn't speak to each other beforehand. We didn't compare notes. We each received our impressions independently, then brought them together afterwards.

And what we found was astonishing.

We both picked up on inflammation in Jasmine's lower back and a distinct sense that she had a urinary tract infection. Erin sensed heat and tightness through Jasmine's spine and tail. I felt an ache in the same region. The verbal expression of the differing non-verbal language we'd received, without hearing each other in advance, was almost identical.

The Truth in the Mirror

That's cross-validation.

And it's been instrumental in growing our confidence.
Because confidence is vital.

Erin and I practiced together regularly, building our confidence as we compared notes and deepened our skills. It was a precious time, not only for learning, but for energetically supporting our two elderly dogs, Annie and Askar, as they journeyed alongside us.

This work takes place against strong headwinds — belief systems, scientific materialism, cultural conditioning.

People who believe there's no such thing as human telepathy are quick to shut this kind of work down. I've experienced it personally. When I felt brave enough to share Jasmine's story with my own family, using her ear infection as an example of intuitive sensing that led to real medical treatment, it was dismissed. Not questioned. Not explored. Just shut down.

In particular, a close family member who works in a medical academic field waved it away without even asking a single question about the experience, let alone showing curiosity about what I was studying, what I had learned, or how it had been verified.

And this is common.

Those of us working with intuitive intelligence don't just face inner doubt; we face external invalidation from the very people who pride themselves on logic. Yet when my dog Askar was recovering from his major surgery, I noticed something quietly written on one of his post-op medication labels: "No data on energetic impact available."

Why?

Because energy, intuitive energy, and emotional energy cannot be measured using current scientific tools.

The Truth in the Mirror

And yet, without energy, we are a decomposing body.
It is half of our physical self.

That's why cross-validation becomes vital.
Because when two non-conferring readers pick up the same information from different input channels, in different ways, it's no longer just intuition. It's shared perception.

When Erin and I read together, we always receive some things in common and some things uniquely.
We don't get the same data. We get complementary data.
We receive nonverbal energy messages in different ways through different clair senses.
We cross-check. We compare tone, language, and feeling.
And again, and again, the messages align.

This is how confidence is built.
Not by convincing others, but by continuing to test what we receive.
By staying curious, staying honest, and tracking what proves itself, again and again.

Ego Projection vs. Clean Data Reception

Let's be honest.
Animal communication, telepathy, and all intuitive arts can sometimes get a bad reputation. Often for good reason.

Not because the skills don't work.
But because people misuse them.

There are those who project their own thoughts or emotions into a reading and call it intuition. There are those who exploit vulnerable clients, especially those grieving a beloved animal or human, to make money or boost their own sense of

The Truth in the Mirror

power. And there are those who, with no training or discernment, decide they're "gifted" and start offering services without ever checking their own inner filters.

This is not integrity.
And it's one of the reasons intuitive work is often ridiculed, feared, or dismissed.

It's also why clean data reception matters so much.

When you tune into an animal, you are not offering advice. You are not diagnosing. You are not channelling your own unmet needs through their voice.

You are listening.

You are reporting, factually, honestly, what comes through.

An animal communicator is not a replacement for a vet.
Let me say that again: we are not diagnosticians.
Our role is to tune into the energetic communication of the animal and share what we receive. That's it.

If we feel pain in the left hind leg? We report it.
If we see an image of a blue ball next to a certain bush? We report it.
If we sense a feeling of being trapped, unseen, lonely, or unwell? We report it.

But we do not tell the owner what to do.
And we certainly don't claim to know more than the animal's own caregivers or medical professionals.

Ethics and congruence are everything.

To be a clear channel, you have to be in a good place yourself.
That doesn't mean perfect.
It means grounded. Self-aware. In integrity.

Energy work is draining.

The Truth in the Mirror

Especially when you're dealing with fear-based or incongruent human energy, which, let's be real, is often the case.

Animals are rarely the "problem."

It's the humans, the owners, who are afraid, shut down, or emotionally unavailable that make the reading heavy.

That is why your own energetic hygiene when working telepathically matters.

You must be able to distinguish between your energy and theirs.

Between a projection and a message.

Between your own emotional history and their present truth.

That discernment is a practice. A discipline.

And it's what separates ethical communicators from those driven by ego or lacking training.

The most powerful readings, the ones that truly serve, come from a place of humility and deep presence.

Not: "I know everything," or "I will tell you something even if it's not real so that you don't think I'm not capable."

But instead: "I'm here to listen. I'm here to receive. I'll tell you what I felt, if anything, and then we'll see what unfolds."

Your Tarot Skepticism — Why You Picked It Up to Help Others Feel Good

I never set out to become someone who read tarot cards.

In fact, I was deeply skeptical of the whole thing.

It seemed... well, totally "woo-woo". Way too weird and "out there".

I'd spent most of my life in high-functioning strategy roles. Data, logic, leadership. That was my currency. Tarot felt too slippery, too performative, too laced with

The Truth in the Mirror

cliché. I'd seen people wield it like a party trick. Others used it to validate their own emotional drama. None of that appealed to me.

But I did notice something else.

People are hurting.
They are searching.
They are desperate for a mirror, something that can hold up a reflection of what they are already feeling but can't quite name.

That's what pulled me in.

I didn't come to tarot because I believed in it.
I came to it because I believed in people.
I believed in the power of being seen, and I wanted to build an online following to share some of my expertise in corporate and life knowledge.

And if a card could help someone feel witnessed, not in a false or fluffy way, but in a grounded, resonant, non-invasive way, then maybe it was worth learning.

I was given a deck. Sent to me by Erin, who had more faith in my intuition than I did.

Quietly.
Reluctantly.
I told myself I'd just experiment with it.
I'd use it the way I use everything: as a tool. Not as truth, but as a potential opening.

And then something remarkable happened.

I started to pull cards for people, quietly, anonymously at first, and the messages landed.
Not generalities. Not vague "this might be your month to shine" nonsense. Specific. Precise. Visceral.

The Truth in the Mirror

People would message me in tears.

"*I didn't tell anyone that.*"

"*How did you know?*"

"*I felt that exact thing just before I saw this.*"

Over and over again.

Tarot, as it turned out, wasn't just theatre.

It was resonance.

It was pattern recognition, soul attunement, subtle intelligence made visible.

And it was one of the most beautiful tools I've ever used to help people feel good about themselves. I run a YouTube channel and help people all over the world understand themselves on a deeper level, building lasting friendships with those who have then chosen to further develop their intuitive skills through my courses and webinars.

I hold a deep respect for the role tarot can play, not as fortune-telling, but as a language. A symbolic mirror. A way to give people a gentle "yes" to what they already know deep down. The design and sequence of cards are a masterclass in psychology, a life journey that all humans experience, a tool created centuries ago that remains resonant today.

And yes, it makes me feel good too, to remind someone they're not broken, not lost, and not alone.

Tarot Case Studies: Precision Insight

I didn't come to tarot because I believed in it.

In fact, I thought it was total nonsense, a "woo-woo" gimmick, an ungrounded distraction. I'd trained my brain in logic, in business, in discernment. But what changed everything was the data. Not blind faith. Not fantasy. Just undeniable results.

The Truth in the Mirror

Initially, I drew one card a day for myself. Over 20 days, the same card appeared 17 times — The Magician. From different decks. Shuffled thoroughly. It shouldn't have been possible.

Then came the case studies.

I invited strangers to email me for free, in-depth readings. No background information. No birth dates, no photos, no astrological data. Just a name. I completed over 50 case study readings in total, some of which were long and some short, and some were for friends, while the majority were for complete strangers. What came through was so accurate, so specific, that it couldn't be dismissed.

Here are just a few examples:

The Red Power Outfit, Hummingbirds, Turtles and Mermaids

One woman's energy came through so clearly that before I even pulled cards, I saw her in a red power suit, bold, corporate, magnetic. I sensed strength and strategy. Then the cards came: An angel of power, two hummingbirds, one looking directly at the person on the card's face. Then, three cards of a turtle and a mermaid appeared in sequence from different decks, all speaking of journeying home.

She messaged back in disbelief.

Red was her power colour. She'd spent years in corporate leadership, helping others make money while feeling drained herself. She was longing for lightness. That very day, a hummingbird had flown directly up to her face, something that had never happened before. And just four weeks earlier, she'd written her first children's book in years. It featured a turtle and a mermaid on a journey to find their home. She sent me a copy of the story; I have witnessed it with my own eyes.

The synchronicity was staggering.

The Truth in the Mirror

The Fairies, the Flowers, and the Land

In another case, I saw a visual, clairvoyant, image of ten fairies dancing around a woman's head. I had no idea if she liked or believed in fairies; however, I felt her deep connection to the land and plants. I used plant-themed card decks, and two different cards from two different decks appeared, both showing the same violet flower. Only three cards in all my decks show that flower, and two came out next to each other in one reading.

Unbeknownst to me, she had just bought a piece of land because it had a natural Fairy Glen, a place she had always felt deeply connected to. She adored fairies. And the day before her reading, one solitary purple flower had bloomed in the glen for the first time. She picked it, placed it in water, and left a crystal in return. She sent me a photograph, and again, I have witnessed the synchronicity with my own eyes.

I wasn't aware of any of this in advance. But the cards and the energy did.

The White Rose, Blue Rose, and the Dog

When I collaborated on a tarot reading with another reader on YouTube, we each laid out symbols for our piles. She used a white rose and a blue rose. I did the same. Dogs were also a core part of both our readings.

Minutes after uploading the video, I clicked on a random pick-a-pile tarot reading from another creator I follow on YouTube. The pile I chose without even glancing at it? It featured a dog charm. Then, shortly into the reading, she pulled three cards in sequence: one with a white rose, one with a dog, and one with a blue rose.

White rose. Dog. Blue rose. In that exact order.

It mirrored our collaboration spreads, something we hadn't published publicly until after the other video went live. A pure, unexplainable echo. Not only that, but the reading also discussed undertaking a collaboration with someone overseas, which was the exact experience we had just been recording.

The Truth in the Mirror

These are not just "nice stories."

They're data points.

I've always been rigorous with how I read. I don't let cards fly randomly. I don't let people tell me their stories. I tune in first. I write or record everything in one flow. And I only read when my own energy is clean and aligned, because that's when the channel is most accurate.

I don't do private tarot readings anymore, they're energetically demanding, and I only have so much to give. But I will never forget the validation they offered, not just for others, but for me.

I started reading tarot not because I believed in it, but because I wanted to reach people, to help them feel seen, understood, and affirmed. I call tarot "lazy-life coaching"; its power to reach people and provide vital self-worth and life skills is profound.

And in doing so, I became exposed to energy exchanges in a way I have no theories to explain.

Tarot is not a crutch.
It's not a trick.
It's a symbolic mirror.

One that speaks, when you're willing to listen.

Tober's Death and the Angels in Plain Sight

The day Tober crossed over, the world stilled.

It was a flat, clear, windless morning, not a single ripple in the air. The vet had opened early for us, a quiet mercy for an old soul making his final journey. My husband (now ex-husband) and I sat in the back of the car with Tober, holding him. Supporting him. There was peace, even in the grief.

The Truth in the Mirror

And then, in the moment he left his body, something extraordinary happened.

Two enormous gusts of wind surged through the stillness, as if the unseen world had opened a door, then closed it again. Tober's spirit moved through. And I felt it. Not in my imagination. Not as a metaphor. I felt the car fill with presence. Jim did too.

I turned to Jim and said, "He's here. He's still here."

We drove home in silence, the presence of Tober's energy feeling vast, even though his inert body was lying wrapped in the back of the car. Our younger dog, Askar, didn't respond to the body. But when we opened the door, he lifted his nose to the wind. Sniffed. Held the air for a moment. Then calmly walked away. He knew.

Later that day, Jim and I went for a bike ride with Askar through the woods. It was one of the few moments of peace and ease we had shared in months. And it didn't last. When we returned, Jim said he was heading to the pub. I didn't want to go, but something nudged me to join him. On the way, a small dog appeared in the road, alone, looking lost.

We stopped. The dog was wary of me, but instantly trusted Jim. For three hours, we tried to find its owner, vet, microchip machine, and waited for the police station to open. It stuck to Jim like glue. I remember looking at it and wondering aloud: "*Tober, did you send this dog to keep us out of the pub? If so, give me a sign.*"

Twenty paces later, I looked down and found a twenty-pence coin on the ground.

It was enough.

We were literally just about to take the dog to the police when a friend appeared at the door, someone who knew the dog and its owner. The reunion was instant. The little guardian had done his job. He had kept us from disconnection on a day when connection had mattered most.

The Truth in the Mirror

That was the beginning of something bigger.

I didn't grow up with angels. I was raised an atheist, science-based, logic-bound, wary of anything mystical. But Tober's passing cracked something open in me. The presence I'd felt that day never left. And not long after, the signs began.

Books and names and conversations arrived without planning, *Angels in My Hair* by Lorna Byrne. Suzanne Giesemann's journey from Navy officer to medium. A conversation with my friend Angela about her baby sister, who had died at birth. I suddenly started receiving energetic impressions from her sister and grandmother, loud and clear. I'd never consciously done human mediumship before. But there it was.

Then one day, something shifted.

I chose to accept that I potentially had a guardian angel. I stopped needing proof and simply surrendered to the possibility. And almost immediately, everything changed.

I asked, gently, quietly, to see my friends one weekend. I hadn't heard from them in days, and I didn't want to chase. I just said to my guardian angel, "*If you exist, and it's meant to happen, please arrange it.*" That night, hours after I would normally expect a message, they texted: *The firepit's on. Come down.*

I went.

We'd built that little firepit spot during lockdown, stones, estuary, peace. We would leave painted pebbles in the wall we built, and people would take them, rarely leaving anything in return. But this time, sitting right on top of the wall, was something that immediately caught my eye...

An angel.

The Truth in the Mirror

Not a symbol. Not a metaphor. A tiny figurine, serene, delicate, completely out of place. I picked it up and sent a photo to Angela, who gasped. "That's the exact angel on my sister's gravestone," she said.

That night, I felt the presence of energy around me like never before. I texted my friend Erin the same picture the next morning. Later that day, she let me know that she, too, had randomly received an angel token. It came in the post, a free gift inside an order from one of her regular jewelery suppliers who had never sent her anything like that before.

Even more energetically aligned and poignant, it turned out to be the last weekend that Askar shared with us on planet Earth. Two days after the firepit, his 15-year-old body gave way, nearly three years after his big sinus-related surgery. He had not made it down to the firepit in weeks because his leg was by then too sore to manage the walk home. The firepit and hanging out with my friends were one of his favourite pastimes, and something made me drive down and share the space with him once more that weekend.

And I wasn't done receiving. I've seen many angel images in the clouds. One time I saw four unmistakable angels aligned in formation, just after I had meditated and received some vivid visualisations about receiving support from a group of angels. I told Erin about it the next day, only to find out that she had purchased four white crystal angels from her crystal wholesaler the previous day.

The day after my mother passed away, I went down to the graveyard where she was to be interred. It was a beautiful day, still, soft, deeply peaceful. I sat down to take it all in. The night before, my dad and I had held her hand at her hospital bedside in her final hours. It had been raw, real, and full of love.

As I sat in silence, I looked up.

There, floating across the sky, was an unmistakable angel-shaped cloud. Clear as day.

The Truth in the Mirror

And beside it — a dragon. Majestic, protective, unmistakable.

Not imagined.

Not hopeful thinking.

Seen.

Something inside me settled. In that moment, I knew she wasn't gone. I knew she was accompanied. I knew something far greater — something intelligent and loving — was holding all of us through the mystery of grief and beyond.

You don't need belief to feel this. You just need openness. Presence. Willingness to receive.

I didn't believe in angels. Now, I have more reasons to believe than not to.

Death Does Not End Communication — It Changes Its Channel

The truth is, none of us really knows what happens when we die.

We can believe, hope, doubt, or dismiss, but unless we've been there ourselves, unless we've touched the edge of it, or felt the unmistakable presence of someone who has crossed over, everything else is just speculation.

Saying there *is* an afterlife is a belief.

Saying there *isn't* an afterlife is also a belief.

The latter often masquerades as fact, but it's not. It's simply the dominant narrative of a culture that has grown cynical of religion and allergic to mystery. Many who reject the idea of life after death do so because they are rejecting the dogma of religion. Fair enough. But throwing out the whole concept of

The Truth in the Mirror

consciousness beyond the body because you don't like how it was once taught, that's not science. That's reaction.

And science, when practiced with integrity, is not in opposition to mystery. It is born of it.

We know that consciousness is real, yet we have no concrete, measurable explanation for where it originates. We know that memory, emotion, and personality can continue in people who've flatlined, who've clinically died, and returned with stories that align across cultures, ages, languages, and belief systems. Near-death experiences (NDEs) are so consistent, so deeply resonant, that dismissing them as "just brain chemistry" is no longer adequate. These aren't hallucinations. They're vivid. Coherent. Often life changing.

People return changed.

They speak of light, of peace, of a presence beyond language. They describe being met by loved ones, by guides, by angels. They describe seeing their lives in panoramic review — not judged but understood.

And here's the thing: many of these people *had no belief* in the afterlife beforehand.

So what are we really dismissing when we scoff at these stories? What does it cost us to consider that maybe, just maybe, we continue on?

We are energy.

That much, at least, is irrefutable. Every cell, every spark of thought, every impulse in the body is electromagnetic. And energy doesn't disappear. It changes form. Why should consciousness be the only thing exempt from that law?

Perhaps what we call "the other side" is not a place, but a frequency. A higher register of energy not held down by the density of physical form, gravity, flesh, pain, resistance. Perhaps those with highly developed clair senses, the intuitive

The Truth in the Mirror

receptors many still call "woo-woo", are simply tuned to receive this frequency. Like a radio picking up a station you didn't know existed.

We don't question sound waves just because we can't see them. We don't call Wi-Fi magic. We've built the tools to measure those things. But we haven't yet built the machines to measure spirit. And until we do, the only way to know is to witness it, with presence, with openness, with a heart willing to learn.

When you've felt it, you can't unknow it.

I have. Again, and again. Through signs, dreams, synchronicities, feelings, and encounters that logic alone cannot explain. And so have countless others.

But here's the caveat: if you refuse to look, you'll never see. If you close your mind, there's no crack through which an alternative truth can shine.

That's why true openness requires being belief-free. Not needing it to be true. Not needing it to be false. Just watching, sensing, listening, and allowing life to show you what it's been trying to all along.

Once you see it, once you feel it, you can't unsee.

And that changes everything.

The Truth in the Mirror

Chapter 6: The Heather Ogilvie Theory of Life Force Transmission

Introducing a replicable energetic model.

The Heather Ogilvie Theory of Life Force Transmission.

My theory suggests that life force, intuition, and psychic information may travel through the body via the electromagnetic fields created by its spiraling structures, and that energy may jump from one coil to another without needing a linear physical pathway.

This could explain why:

- You feel something in your gut before your brain registers it.
- You sense someone's emotional state without a word being spoken.
- A memory is "held" in the hands, the belly, or the heart, long after the mind has forgotten.
- Information, inspiration, or insight arrives as a "download" from seemingly nowhere but is felt in your body before it becomes thought.

It may even explain why life force, the animating intelligence we can't quite define, stays in the body at all. Why energy doesn't just disperse the moment it's created. Why it remains *coherent, contained,* and *connected,* until something shifts, and it leaves.

This theory is not a replacement for existing scientific models. It is a *possibility.* A bridge between energetic sensitivity and biological structure. A call to look deeper, not just into what we feel, but why we feel it, and how that feeling might be moving through us in unseen ways.

This chapter is here to ask a question: *"How is energy travelling through you in the first place?"*

The Truth in the Mirror

What if your body were a highly tuned receiver, a resonant coil, holding a field so intelligent, so precise, that it can carry life force, insight, and connection across time, space, and species?

Life Force Travels Through Biological Spirals Like Transformers

This is my theory. It may not be written in a textbook — yet. But I have lived it, tracked it, and felt it more times than I can count. It's the foundation of everything I now understand about energy, intuition, and the invisible threads that connect us to each other and the world around us.

Life force is not just a mystical current. It is a measurable, trackable energetic force that moves through biological structures we already know exist, especially spirals. In fact, this theory is not based on the esoteric, but on resonance, structure, and lived experience. I call it *The Heather Ogilvie Theory of Life Force Transmission*.

I believe life force does not simply exist in us. It moves through us. It travels through the coils, spirals, and currents of our physical and energetic bodies, the way electricity flows through a transformer. Transformers operate on a simple principle: energy enters a coil, builds momentum through the spiral, and then jumps to the next conductive structure. More spirals. More jumps. More power. What if the same is true for us?

Double helix DNA. The loops of the gut. The folds of the brain. The spirals in your fingertips. These are not just anatomical features; they are conductors. In physics, spirals and coils are used to conduct, store, and transmit energy. They amplify and transform the current. That's why you'll find spirals in everything from radios to transformers to wireless chargers.

So what if the body operates in a similar way? What if your DNA isn't just storing genetic information, but also conducting and translating energetic data, yours and others'? What if each coil in your gut or loop in your brain is acting like a biological

The Truth in the Mirror

transformer, amplifying intuitive input, emotional signals, even the unspoken language of animals or people thousands of miles away?

I believe it does. I believe your body is more than a vessel. It's a system, an intelligent, highly tuned instrument, built for sensing, translating, and transmitting life force. And when all the spirals are resonating in harmony, when the gut, the brain, the heart, the fingertips, the fascia, the field are aligned, they start to sing. That's when we experience the profound clarity of intuition. That's when we feel truth. That's when we hear what isn't spoken and know what hasn't yet been said.

Every time I sit down to tune into an animal, a person, or a field of possibility, the energy often moves into my body before landing in my mind as a fully formed verbalised output. That's not random. That's resonance.

And just like electricity jumps between circuits, energy jumps through us. From coil to coil, from field to field, via internal structural conductors. Some look to chakras, energy centres often depicted as spirals, which may explain their persistent presence across cultures. But we don't all need to subscribe to chakra models to grasp this, and I theorise that energy starts in even smaller coils within the body than our chakras, namely our DNA. Energy jumps, spirals amplify, and the body is more than a sack of organs. It's a resonant field.

This theory didn't come from books or labs. It came from direct, repeated, validated experience. I have felt the currents of animals who were thousands of miles away. I have received information that was later verified with astonishing specificity. I have tuned into images, emotions, smells, tastes, and physical sensations that did not originate in my own body but were undeniably real. And I am not alone in this. My body is not special. I have trained with people and trained people. We all possess the same capability, so it must exist somewhere within our bodies.

How could this be happening? Not through imagination. Not through magic. But through structure.

The Truth in the Mirror

Our bodies are exquisitely designed instruments of energy transfer. The spirals in our biology are the equivalent of finely-tuned antennae, wired to pick up and relay signals from the world around us, and beyond.

This is not "woo-woo". This is resonance, frequency, and field.

Life force doesn't need to be believed in. It needs to be understood. Measured. Explored.

And one day soon, I hope to collaborate with researchers, physicists, and intuitive practitioners who are willing to test this theory. To measure energy as it moves. To map the resonance and energetic fingerprint of DNA under emotional stimuli. To explore the electric intelligence of the gut, the sensitivity of the fingertips, and the vibrational echoes of the heart field.

Because this is not mysticism disguised as science. This is science finally listening to what the body has been saying all along.

Intelligent energy is already alive in every single one of us.

Energy Jumps — Not Chakras, But Conductors

We've been taught to think of energy moving in tidy vertical lines, up the spine, through seven fixed chakras, from root to crown. And while that framework has its value (and resonance for many), it doesn't tell the whole story.

In fact, most of what I experience in intuitive work doesn't feel linear at all.

It jumps.

It flickers through my gut or tightens my chest. It tingles in my fingertips, or lands behind my eyes as a sudden image, or in my brain as unquestionable knowing. It loops. It spirals. It dances. Sometimes it floods every cell at once.

This isn't just a metaphor.

The Truth in the Mirror

In physics, we already know that energy doesn't need a physical connection to travel from one place to another. Coiled systems, such as those found in transformers or wireless chargers, enable electromagnetic energy to leap across gaps. That's how resonance works: when two systems are tuned to similar frequencies, energy can transfer between them, seemingly without contact.

This is the basis for my theory.

I believe that when energy "jumps" inside the body, it's moving between spiral structures that are naturally tuned to each other, gut to brain, brain to heart, cell to cell. This transfer doesn't require a central pipeline. It needs resonance.

Which is why the chakra model, while symbolically helpful, is incomplete.

Chakras are often depicted as spinning wheels or spirals, and that's where they align with what I'm proposing. The spiral is the key. Not the location or the colour or the hierarchy, but the form. The spiral is what allows energy to build, amplify, and jump.

And that means you don't need to focus on "opening your third eye" or "clearing your sacral." You need to tune your coils. Harmonise your spirals. Ground your resonance. Unclutter your unique frequency.

Because this is not a vertical hierarchy of seven lights.

This is a multidimensional field of electromagnetic intelligence, moving, leaping, living inside you.

It's not about belief.

It's about conduction.

The Truth in the Mirror

Sensory Sensitivity as Data Acquisition, Not Mysticism

Let's get one thing straight: being energetically sensitive is not a defect, a disorder, or a personality type.

It's data acquisition.

It's your body doing what it was built to do, receive, translate, and respond to frequency.

We don't question a satellite dish for picking up signals. We don't call a radio "too emotional" because it's tuned to a wide band of transmissions. And yet, when a human being picks up subtle information, a mood, a knowing, an image that proves accurate, we're labelled "overly sensitive," "too intense," or "woo-woo."

Enough of that.

Sensory sensitivity is not about fragility. It's about bandwidth.

The more sensitive your system is, the more you pick up. And when you learn to translate that information, rather than be overwhelmed by it, it becomes intelligence. Not just emotional intelligence, but energetic intelligence. Bodily intelligence. Environmental intelligence.

This is what intuitive development really is: not a mystical elevation, but a fine-tuning of your innate sensory apparatus. An awareness of what's yours, what's not, and how to interpret the energetic signatures moving through your field.

You might feel a chill in your arms before someone texts. You might get a sour taste in your mouth when an animal is unwell. You might have a vivid dream and receive confirmation the next day.

That isn't magic. It's a form of attunement.

The Truth in the Mirror

It's your gut, your vagus nerve, your heart field, your DNA coils, all receiving input before the brain can interpret it consciously.

If anything, this is where science needs to catch up. Because the evidence isn't just anecdotal. It's embodied. Repeatable. Verifiable through lived experience and cross-validation, as we've already shown in earlier chapters.

This is how intuition really works. Not by bypassing the body, but by decoding the data it's already receiving.

You don't need to become more magical.

You just need to learn to listen.

Psychic Awareness Framed as Pattern Recognition

Psychic ability has often been cloaked in mystery, as if it belongs only to a gifted few or operates outside the laws of nature.

But what if it's simply advanced pattern recognition?

This doesn't mean it's not extraordinary. It is. But not because it's mystical, because it's precise.

The brain is a pattern-seeking machine. It scans for associations, notices rhythms, makes predictions based on repeated inputs. That's how we learn language, read faces, navigate relationships, and survive. It's not "out there." It's already built in.

Now expand that beyond the five physical senses.

Imagine your entire energy field is doing the same thing. Picking up vibrational cues. Matching current frequencies to previous ones. Alerting you through images, feelings, tastes, smells, flashes of knowing, long before your conscious mind understands what's happening.

The Truth in the Mirror

This is how I experience psychic awareness.

It's not just about "seeing the future" or "talking to spirits." It's about recognising energetic patterns. Subtle shifts. Recurring signatures. It's about tuning into the repeated hum of a truth that lives just beneath the surface of the visible world.

That dog's energy feels vibrant or dull.

That person's frequency matches someone I've felt before, someone who was carrying grief they hadn't voiced.

That land holds a vibration I've known in dreams, in visions, in the body of a horse who showed me the storm before it came.

That house doesn't feel right for me to live in; I don't know why. I can't explain it, but it just doesn't feel right.

This is not imagination. It's recognition.

And it's not confined to those who call themselves psychic. Anyone can develop it, because it starts with presence. With attention. By letting your system notice what it already knows.

Once you've seen these patterns, once you've felt them enough to trust them, they become maps. They guide you. And you stop asking, "*How do I know this?*" or even worse, dismiss the message altogether, and start asking, "*What is this trying to show me?*"

Because the moment you trust that awareness, it begins to sharpen. To clarify. To lead.

And life starts making more sense than it ever has before.

The Truth in the Mirror

The Body as Transmitter, Receiver, and Amplifier

Your body is not just a passive shell.

It's not a container for consciousness. It *is* consciousness, in form, in motion, in vibration.

We've been trained to think of intuition as something ethereal, floating "outside" or "above" us. But what if the most potent psychic tool you have is not your third eye — it's your nervous system? Your gut? Your fascia? Your fingertips?

In the Heather Ogilvie Theory of Life Force Transmission, we've already explored how spirals, such as DNA, intestines, and neural loops, behave like transformers, allowing life force and energy to jump, build, and transmit. But let's ground this even further.

Your body is doing three things all the time:

1. Transmitting – sending out a frequency, a signal, an intention, a vibration.

2. Receiving – picking up signals, mostly subconsciously, from other beings, environments, thoughts, and fields.

3. Amplifying – increasing the volume of certain signals through resonance, emotion, coherence, or attention.

When you walk into a room and instantly feel a shift — that's your body receiving. When you think of someone and they message you five minutes later — that's your frequency transmitting.
When you sit with an animal, and both of you go still, synced, deeply attuned — that's mutual amplification.

And when you draw cards, as I have done hundreds of times in intuitive tarot readings, you feel the energy move through you. The moment your hand touches

The Truth in the Mirror

the deck, something happens. The spirals in your fingertips, the vibration in your gut, even perhaps the DNA embedded in the paper itself, they connect. They resonate.

I've often wondered if the DNA in the cards, the paper, the ink, the touch, plays a role in the energy that flows through a reading. Because something is being transmitted. I feel it. I receive images, impressions, and full-body sensations before I've even turned a single card. And I create verifiable, accurate readings for total strangers. That's not imagination. That's interaction.

I don't read live in front of people because the presence of another energy field can disrupt my tuning. I record readings in solitude, where I can attune to the frequency of the person's or collective's field without interference, where my body can act as the clear channel it was designed to be.

I theorise that the spirals and loops of matter in our brain and gut are part of the energetic transference system. They don't just conduct energy. They *amplify* it.

Think of it like this: if many small DNA coils live inside a larger spiral, say, the loop of the intestines or the folds of the brain, the larger spiral acts like a megaphone. It amplifies whatever energy is being transmitted or received. That's why I believe animals like elephants, whales, and horses are so attuned, their vast, complex guts may serve as enormous amplifiers of energetic information.

They don't just sense danger. They sense intention. Emotion. Frequency.

So do we.

It's just that we've forgotten how to listen. Instead, our egos are lost in constant transmissions of pain, lack, emotional disconnect, mental anguish, needing to be seen, analysing, pleasing, sacrificing, fearing, and controlling.

But the body remembers. It always has.

The Truth in the Mirror

Your spirals are singing. Your gut is listening. Your fingertips are feeling. You are already transmitting and receiving energy at every moment, sometimes across rooms, sometimes across continents.

It's not magic. It's resonance.

And when we come into alignment, when our systems sing in coherence, we don't just "get intuitive hits." We become the instrument.

Vision for Collaborative Research and Model Refinement

This theory isn't meant to live in isolation.

The Heather Ogilvie Theory of Life Force Transmission is not a declaration of absolute truth — it's an invitation. A hypothesis. A conceptual framework built from lived experience, subtle perception, and years of direct evidence through intuitive, psychic, and animal telepathy work.

But for it to grow, it needs dialogue.

I believe there is an opportunity here, one that bridges ancient knowing with emerging science, intuitive insight with measurable data. The spirals that transmit energy through our bodies aren't abstract. They are biological structures. They exist. They can be measured. And I believe we are on the edge of developing the tools to do just that.

I would love to collaborate with physicists, neuroscientists, biologists, intuitive practitioners, and sensitive technologists to explore this further. To track:

- How energy moves between biological spirals
- Whether certain individuals have stronger energetic fields or receptor sensitivity
- How emotional states affect electromagnetic coherence in the gut and brain
- What role DNA plays in non-verbal, energetic transmission

The Truth in the Mirror

- How energy signatures are amplified through gut or cardiac resonance
- Whether intuitive accuracy correlates with measurable physiological shifts

Imagine building technology that can map resonance in real time. Not to validate intuition for the sake of "proving" it, but to understand how it works — biologically, energetically, and experientially.

We've already begun exploring quantum entanglement, heart field magnetism, and brainwave entrainment. Why not this?

What if animal telepathy, intuitive hits, even psychic dreams aren't anomalies, but the normal function of a well-tuned, resonant and cohesive set of biological coils?

What if life force can one day be mapped?

What if the soul — the energetic presence of a being — is measurable, not in mass or weight, but in coherence, light, sound, or frequency?

And what if everything we thought was mystical is actually mechanical, not in the cold, lifeless sense of machinery, but in the precise, elegant dance of form and field?

This is the frontier, I believe, we are entering.

Not the age of artificial intelligence.

But the age of *energetic intelligence*.

And it lives in you. In your spirals. In your senses. In your gut. In your fingertips. In your knowing.

We've only just begun to listen.

Part Three: Applications & Implications

A New Era of Human Intelligence

The Truth in the Mirror

Chapter 7: Intuition Is Not Optional

Why this matters now more than ever.

Intuition Supports Leadership Under Uncertainty

The world we live in is moving faster than our existing systems can keep up with.

Linear models of logic, strategy, and prediction are crumbling in the face of rapid change, emotional complexity, and global uncertainty. From climate disruption to financial volatility, technological acceleration to cultural collapse, mental health issues to spiritual disconnect, the leaders of today are no longer being asked to "decide well."

They're being asked to *feel wisely.*

This is where intuition comes in. Not as a soft skill. Not as a mystical nice-to-have. But as an essential form of intelligence, *biological, energetic, and vital to leading in times of chaos.*

In my work with founders, CEOs, and mission-led entrepreneurs across the world, one truth has emerged again and again: logic can only take you so far. It's your capacity to read the room, feel the future, and sense the unseen that defines whether your leadership will land, or fracture.

This is why I created Leading Beyond Logic, visionary training to support leaders in shifting from dominance to presence, from control to coherence, from ego to soul. It's about cultivating the inner sensing system that lets you make decisions *in real-time,* even when the data is incomplete, and the stakes are high.

Because the truth is, the leaders who will thrive in the next decade aren't the ones who know the most.

They're the ones who can listen the deepest.

The Truth in the Mirror

This isn't just about "gut instinct" or "intuition." It's about *energetic intelligence*.

The ability to access the full spectrum of information available — cognitive, emotional, relational, and subtle. The leaders who can feel the undercurrents, sense the timing, and stay in resonance with their own integrity will be the ones who build trust, hold clarity, and create futures worth walking into.

Intuition is not a replacement for strategy.

But it is a requirement for wisdom.

And without it, leadership becomes blind, brittle, and disconnected — a reactive performance, instead of a grounded act of service.

Decision-Making Improves with Subtle Data Awareness

We often think of data as numbers on a spreadsheet or feedback from a survey. Tangible. Measurable. External. But the truth is, some of the most powerful data we ever receive doesn't come through those channels. It comes through the body.

A tightening in the chest when something's off.
A quiet drop into calm when something is true.
A flash of vision. A download of knowing. A feeling in your stomach that won't go away.

This is *subtle data*. And learning to work with it changes everything.

It's not about overriding logic. It's about enriching it.

In intuitive leadership, we're not trading numbers for feelings. We're *adding layers of intelligence* to the decision-making process — layers that include emotion, energy, timing, and resonance.

The Truth in the Mirror

Most leaders already do this; they just don't name it. They call it "gut instinct" or "reading the room." But what if we stopped treating this as a lucky accident and started treating it as a *trainable skill?*

Subtle energy awareness is the difference between pushing a decision because it looks good on paper or would provide short-term gains to shareholders and pausing long enough to feel that something isn't aligned.

It's the difference between a rushed yes and a resonant no.

It's what lets you *hear what isn't being said,* in meetings, in negotiations, in moments where words are present, but truth is not.

This is especially crucial in volatile contexts, where the stakes are high, the path is unclear, and the answers don't live in a playbook. The more subtle energetic data you're attuned to, the better you can navigate ambiguity without collapse.

And this isn't guesswork.
This is embodiment.
This is how real leaders lead.

The Power and Presence of Subtle Energy

Why You Ignore It at Your Own Cost

Subtle energy is the most prevalent, yet least acknowledged, force in our modern lives. It is the invisible thread that runs through everything, the field through which intuition travels, the backdrop of every emotion, decision, and interaction. It is not just "around us." It's *in* us. It's *us.*

This energy, sometimes called life force, prana, chi, spirit, or just atmosphere, shapes what we feel before we understand. It's what gives a room its mood. A conversation its charge. A landscape its magic. It doesn't need to be believed in to affect you. It already does. Every day.

The Truth in the Mirror

When people hear the word energy, they often think of electricity, adrenaline, or "good vibes." But subtle energy is something more precise. It is the unseen field in which all things are connected, shaped, and guided. It is the quiet architecture behind the material world, the intelligence that we sense but don't always know how to name.

It is not mystical fluff.
It is not pseudo-spiritual performance.
It is the fabric of everything.

Subtle energy flows through and around all living beings. It's the language of the soul, the current of intuition, and the field in which true perception occurs. Animals live by it. Nature moves by it. And whether we're aware of it or not, humans are constantly influenced by it.

You've felt it.

It's the shift in the air before someone walks into the room.
The internal contraction when you hear something that isn't quite true.
The expansion in your chest when someone speaks from the heart.
The sudden thought to change lanes, right before an accident would have happened.

These are not random flukes. They are real-time communications through a subtle network. A system that transcends logic and time but does not bypass your body.

Your body is an intelligent instrument of subtle energy. It picks up, translates, and responds to frequencies faster than the conscious mind can compute. This is why intuitive information so often arrives before explanation. It's felt as a knowing, a gut hit, a pull. And it's also why ignoring it comes at a cost.

Subtle energy is not abstract philosophy. It is embodied, measurable in its impact, and shaped by everything from the food we eat to the thoughts we think. It influences our choices, creativity, relationships, and leadership. Here's how:

The Truth in the Mirror

Self-Awareness & the Energy of Perspective

Your energy field is shaped by your perspective on the world. If you believe life is dangerous, people are selfish, or you're not good enough, your energetic field shrinks. It hardens. You stop receiving. But when you choose self-awareness — when you pause, question your thoughts, and align with truth — your energy opens. Insight flows. Intuition sharpens.

Nature as Energetic Recalibration

Walk into a forest and your nervous system shifts. Stand beside the ocean and your breath deepens. Nature carries frequencies that naturally harmonise the human field. That's not metaphor. It's electromagnetic coherence. Subtle energy responds to nature and is healed by it. That's why animals live so intuitively: they are in constant resonance with the Earth.

Human Environments & Technology

Our homes, offices, and cities shape our energy. Fluorescent lights, electromagnetic fields, cluttered spaces, and constant noise all disrupt the subtle systems through which intuition flows. Technology itself isn't bad, but how we use it matters. If you are constantly absorbing chaotic frequencies without taking time to reset, your intuitive signal becomes scrambled. Like trying to hear a whisper in a hurricane.

Media & the Energetic Narrative

Every story you watch, read, or scroll carries an energy field. News saturated with fear shrinks your awareness. Social media echo chambers distort your resonance. We are not just consuming information, we are embodying it. If you want clearer intuition, guard your inputs like your life depends on them. Because energetically, it does.

The Truth in the Mirror

Money, Self-Worth & Energetic Congruence

Money is a value exchange, a tool for marking timing differences when you swap your creative energy in return for someone else's creative energy. It is not just a number. When your self-worth is low, you energetically block abundance. When you give from depletion or receive without gratitude, energy becomes distorted. But when you act from integrity, value your gifts, and exchange with consciousness, your financial field expands. Intuition, by extension, becomes sharper because your field is clean.

Food, Movement & Physical Sensitivity

The food you eat carries vibration. Highly processed, chemically altered foods dull your system. Living, fresh, plant-based foods heighten it. Same with movement. Exercise clears energetic debris. It reactivates life force. When your body is energised and clean, your intuitive channels open. Your gut, quite literally, speaks louder.

Laws, Societies & Structural Energy

Every system, governments, legal codes, education, carries energy. If it suppresses truth, intuition dims. If it fosters fear, your nervous system constricts. We live inside energetic scaffolding built by belief. The more conscious the structure, the more naturally our intuitive intelligence can thrive. Intuition is not anarchic, it wants coherence. But it cannot survive in systems built on fear.

Our Bodies & Health as Energy Literacy

Pain, fatigue, illness — these are energetic messages. Your body doesn't lie. When your energy is aligned, health flows. When it's blocked, symptoms appear. You don't need to *believe* in subtle energy to feel that tension in your chest when something's wrong. Or the lift in your heart when something's right. The body is

The Truth in the Mirror

a translator. The wiser you become in its language, the clearer your intuitive signals.

Love, Relationships & Connection

Love is not just a feeling — it is a frequency. The most powerful one. When you're connected to someone who sees and honours you, your energy field expands. You become more of yourself. When you're around someone who diminishes, manipulates, or confuses you, you contract. Partnerships, family dynamics, even fleeting interactions with strangers — all of these live in the energetic realm before they take form in words.

Purpose, Work & Creative Flow

Your purpose lives in your energy field long before it appears in your job title. It pulls at you. It whispers. It sends signs. The same is true for creativity. Creative blocks are often energetic blockages. When your system is aligned, when you are in integrity with yourself, intuition becomes your compass. Work becomes contribution. Passion flows. Not from effort, but from resonance.

Adventure, Fun, and Play

Joy isn't an indulgence. It's an energetic state that fuels intuition. When you play, laugh, and explore, your field becomes more receptive. Downloads arrive. Insight flows. You reconnect to source. A life without pleasure is a life where energy stops moving. Adventure is not a distraction from your path. It's part of the calibration.

This is subtle energy.
It's real. It's relational. And it's always speaking.

If we want to lead, love, or live well in the world ahead, we must learn to hear it.

And that begins with one decision:
To stop dismissing what we *feel* simply because we can't *prove* it.

The Truth in the Mirror

The Cost of Ignoring This Inner System

What Happens When We Lead Without It?

We've built a world that reveres intellect but neglects intuition.
That quantifies, but rarely questions.
That tracks key performance indicators but ignores gut indicators.

And we are paying the price.

When we ignore subtle energy, we disconnect from truth. Not abstract truth, *actual* truth. The kind that tells you something's off before it breaks. The kind that warns you about a person before the evidence arrives. The kind that taps you on the shoulder in a boardroom and whispers: This will cost more than money.

Leadership that dismisses intuition makes decisions that *look* good but *feel* wrong.
And eventually, the feeling wins.
Because your nervous system always knows before the data can catch up.

We see this everywhere:

- Burnout from ignoring the body's early signals
- Personal misalignment from chasing what "should" matter instead of what actually does
- Business decisions that fail because they ignored human energy, timing, or resonance
- Relationships that collapse because something was off, but no one listened to their own knowing

This is the real cost:
A life, a business, a body, a culture, a world, completely out of sync with itself.

And the longer you deny that energy is a factor, the more energy it takes to fix what goes wrong.

The Truth in the Mirror

But here's the truth:
Your system wants to bring you home.
It sends signs. It gives you physical nudges.
It contracts around dishonesty and expands around alignment.

When you ignore this inner system, you get noise.
When you learn to trust it, you get clarity.

The future belongs to those who are brave enough to combine data with discernment.
Who are willing to lead not just with spreadsheets, but with soul.

Subtle energy also helps explain what we often refer to as "downloads", moments of sudden insight or clarity that seem to come from nowhere. You ask a question. You go quiet. And then, out of the silence, something lands. A sentence. A strategy. A vision. A no. A yes. An image so precise it takes your breath away.

That's subtle energy at work, delivering truth through resonance.

It is shaped by your thoughts, your field, and your intention. But it is not limited by them. Subtle energy responds to coherence. The clearer you are, the cleaner your channel, the more accurately it flows.

This is why presence matters. Why energy hygiene matters. Why discernment is everything.

Your energy field expands or contracts depending on what you consume, how you rest, how you feel, and who you surround yourself with. Nature enhances it. Screens distort it. Congruent people nourish it. Energy vampires drain it.

And just like light through a lens, your clarity determines how sharply you can "see."

The Truth in the Mirror

The modern world teaches us to trust logic. But logic is just one part of intelligence. Subtle energy offers another. It's the field behind the field, the place from which everything arises, and the place into which everything returns.

Once you begin to work with it, consciously, consistently, your entire experience changes.

You no longer lead from fear.
You no longer guess.
You listen.
You attune.
You act from a deeper kind of certainty, the kind that doesn't shout, but sings.

This is subtle energy. And learning to work with it isn't optional. Not anymore.

Reclaiming Self-Trust in a Noisy World

The truth is not hard to find when you trust your intuition, it's the space to hear it that is.

We live in an attention economy. Everyone is trying to sell you something: a product, a belief, a headline, a version of yourself that's more desirable, more profitable, more palatable. Algorithms reward sensationalism and outrage. Culture rewards performance. And the result?
We've become more disconnected from our own inner signals than ever before.

This disconnection isn't accidental.
It's cultural. Systemic.
And it's especially pervasive in those of us who were trained to achieve, to lead, to get it right.

You may have been told your body was too sensitive.
That your feelings were irrational.

The Truth in the Mirror

That your insights needed evidence.
That your inner knowing didn't count unless someone else validated it.

But here's the truth:
There is no clarity in a world that runs on noise.
The only clarity you'll find is in the calm you claim for yourself.

And that clarity comes not from consensus, but from *resonance*.

Reclaiming self-trust means turning inward again.
It means becoming exquisitely attuned to your own system — how it says yes, how it says no, how it signals danger, how it affirms truth.
It means remembering that intuition is not a mood or a magic trick — it's your deepest inner compass, grounded in biology and reinforced through practice.

It means shifting from asking,

"What should I do?"
to
"What do I know in my gut?"
and
"What does my body feel when I say yes to this?"

This level of self-trust is not arrogance.
It's integrity.
It's what allows you to navigate uncertainty without panicking, to lead without dominating, to hold vision without needing constant reassurance.

And it's what the world is starved for.

We are not here to build more performance-based, attention-seeking leadership. We are here to build congruent, soul-led, quietly powerful presence. The kind of presence that listens first, not for consensus, but for alignment.

When you stop outsourcing your authority, you become available to truth.

The Truth in the Mirror

You stop chasing answers. You stop bending to feedback loops that don't serve you. You stop giving your energy to people or systems who don't honour it.

Instead, you become the field.
The grounded one.
The lighthouse, not the weathervane.

And from this place, your decisions shift.
Not from reactivity or fear, but from deep congruence.

Because when you trust yourself, you move differently.
You speak more clearly.
You pause before responding.
You don't need the room to agree — only your body to resonate.

Self-trust is not instant. It's a practice. It's rebuilt through consciously choosing to reframe every small moment of the day, selecting quiet over performance, integrity over approval, and truth over comfort.

But once it's restored, there's no going back.

A New Kind of Intelligence for a Volatile Age

Subtle perception is the next frontier of human evolution. We are living in a time that asks more from us than ever before — not just intellectually, but energetically.

Old forms of intelligence are no longer sufficient.

IQ alone cannot solve the crises we face. Logic doesn't adapt fast enough to complexity. And traditional models of "knowing", ones that demand hard evidence before action, can leave us paralysed in a world that's already moving.

The next evolution of human intelligence is not more data.
It's deeper perception.

The Truth in the Mirror

Intuition, when grounded in the body and honed with practice, becomes a precision tool in chaotic environments. It allows us to respond in real time to subtle signals. It helps us notice what others miss. And it gives us access to a broader field of data, one that includes emotional truth, energetic resonance, and pre-conscious awareness.

This isn't soft. It's strategic. It's not about being "psychic." It's about being present. Fully, viscerally, attuned to your body, your field, your team, your mission, your environment.

When we talk about a new kind of intelligence, we are not talking about replacing logic. We're talking about expanding it. Integrating it with the kinds of knowing that live in the gut, the heart, the skin, the senses that react faster than thought.

This is the intelligence that feels the tension in a room before it's spoken.
That senses the moment to pause, shift, or act.
That catches what others don't say — but what their energy screams.

This is the kind of intelligence that:

- Helps leaders navigate moral complexity with integrity
- Allows healers to sense what isn't yet visible
- Guides creatives to the edge of innovation
- And gives everyday people the power to discern what's true for them, even when the world is screaming otherwise

It's not mystical. It's measurable, even if we haven't yet built the tools. The body is a receiver. The heart is a frequency field. The gut is a neural network.
This kind of intelligence is already in you. It's just waiting to be trusted.

And as the world becomes more volatile, more uncertain, noisier, and more destabilising, this form of intelligence becomes not just helpful, but essential.

Because those who can hear the subtle will move first.
And those who lead from resonance will build what's next.

The Truth in the Mirror

Chapter 8: The Science We're Missing

Where the research must go next.

Open Call to the Scientists of the Future

To Neuroscientists, Physicists, Veterinarians, Biologists, Medics, Psychiatrists, Energy Experts, and Everyone In-Between. What if the reason we haven't been able to prove the reality of intuition is not because it isn't real, but because we haven't built the right instruments yet?

We're standing on the edge of a new understanding. Not mystical. Not abstract. But rooted in biology, physics, resonance, and systems.

Intuition, telepathy, and psychic senses are all sensory functions — not a fantasy — and it's time to bring them into the laboratory. Not to dissect them, but to honour them. To decode them. To trace the signal path through the body and learn where it's coming from.

This is a bold, public, heartfelt invitation to the scientific community. To those who know that true discovery lives not in defending old dogma, but in asking better questions.

We need you. All of you.

- **Neuroscientists** to track what happens in the brain during moments of intuitive "knowing", including altered states, alpha/theta brainwaves, and right-hemisphere dominance during psychic downloads.

- **EEG and EMG researchers** to observe micro-signals across the nervous system during live communication with animals, people, or energetic fields.

The Truth in the Mirror

- **Cardiologists and heart coherence experts** to map the field dynamics of the heart when emotion is exchanged silently between species, especially across distance.

- **Gut-brain axis researchers** to investigate how messages from the enteric nervous system precede cognition, especially in instinctive and emotionally charged interactions.

- **Quantum physicists and entanglement theorists** to explore the nature of non-local awareness, when two beings feel each other without conventional contact.

- **Biophysicists** to study the structural intelligence of DNA and its potential function as a frequency conductor, not just a genetic archive.

- **Veterinarians** and **animal behaviourists** to rigorously test and record accurate telepathic communication between humans and animals across species.

- **Medical doctors, GPs, and psychiatrists** to reopen the conversation about what constitutes "real perception", and how inner knowing often precedes diagnosis.

- **Systems theorists** to model the body, field, and environmental feedback loops that govern human energetics, intuition, and coherence.

- **Electromagnetic researchers** to trace the subtle, low-frequency shifts in field strength and polarity during intuitive events.

- **Energy technologists** to create the next generation of wearable or ambient sensors, devices that go beyond heart rate and movement and

The Truth in the Mirror

start to capture the human field itself.

- **Educators** to explore how children learn and respond differently in coherent energetic environments, or when their own energy fields are validated rather than dismissed.

- **Linguists and cognitive scientists** to map how intuitive data "lands" in the psyche, through image, word, feeling, or metaphor, and what this reveals about the structure of consciousness itself.

We need open minds. Cross-disciplinary collaboration. And the humility to recognise that what we don't yet understand is not "woo-woo" or weird — it's just waiting for the right lens.

Because we have enough lived evidence to know this is not imagination.

It is not wishful thinking.

It is repeatable. It can be validated. And often verifiable across multiple sources, even when those sources have no knowledge of each other.

Intuitive communication and telepathy are data sets we've been taught to ignore. But data doesn't stop being valid just because it's subtle.

So I'm inviting you, not just to study this, but to pioneer it.

Imagine being part of the first wave of peer-reviewed, grounded, bio-energetic science that redefines what it means to be human.

Imagine helping the world understand that emotional resonance, energetic perception, deep interspecies connection, and non-human intelligence are not fringe experiences — they are foundational truths of biology.

The future of science is integrative. And we need you on the team.

The Truth in the Mirror

A Plea for EEGs, Frequency Studies, and Gut-Brain Scanning

If you are open-minded enough to accept that intuitive awareness is real, that it can lead to verifiable outcomes, cross-validated experiences, and specific insights not obtainable through logic alone, then also accept this:

We are overdue for the technology to catch up.

Because the body is speaking. The question is: Are we ready to listen?

For too long, subtle energy, intuition, and psychic perception have been relegated to the realm of unmeasurable science, not because they are unscientific, but because we have lacked the courage — and often the funding — to study them properly.

But now we are ready.

And here's where we must begin:

We need EEG studies on intuitive states.

To map what happens in the brain and the rest of the body when someone receives an intuitive hit. What are the patterns of brainwave, gut, and heart activity when a person correctly describes an animal's emotional state, pinpoints the location of a missing pet, or receives a spontaneous image that later proves accurate?

Are we seeing increased theta wave coherence? Are specific hemispheres of the brain more active? Does the brain enter synchrony with another's signal field? What is going on energetically in the rest of the full body system?

We will never know unless we test.

The Truth in the Mirror

We need frequency analysis of the body's electromagnetic field during intuitive events.

This includes the toroidal field of the heart, the pulsing electromagnetic field of the gut, the fine variations in skin conductivity, breath rhythm, and body temperature. What about the human aura, the energy that extends beyond the body? When does external energy first pass into our own auric field system? Subtle shifts are likely happening, we simply haven't been measuring them with the right sensitivity.

We don't need to start with complex machines. We need more sensitive ones. Real-time frequency mapping of intuitive states, especially those verified after the fact, would be a breakthrough.

We need gut-brain-heart axis scans during energetic exchange.

What if the gut knows before the brain responds? What if the first signal from the animal or field of consciousness doesn't land in the mind, but in the stomach?

Are there biomarkers of intuitive reception in the microbiome?
Can we see energetic "lighting up" in the gut's neural network prior to conscious awareness?

There's emerging evidence that trauma, emotion, and memory are stored in the gut, so why not intuitive data too?

We need pattern recognition tools trained on intuitive speech and sensation reports.
How do highly intuitive people describe their experience?
Can AI models help identify the structures and linguistic patterns that distinguish genuine intuitive data from imaginative conjecture?

This isn't about policing intuition or dismissing telepathy because you can't validate it, it's about illuminating its structure and creating systems to measure it.

The Truth in the Mirror

The same way we train computers to identify cardiac anomalies, we could train them to identify linguistic or emotional markers of a true energetic signal.

We need to test the fields between beings.

As energy is shared between humans and animals, humans and humans, and humans and the landscape, it must be traceable. Not by eye, but by field strength, by resonance, by coherence.

Can we measure what changes when two beings emotionally attune?

Do we see matched frequencies in the heart field? Do we observe shifts in magnetic alignment or field pulsation when connection is formed, or when it's broken?

This is where the future is heading.

It's not enough anymore to say, "*This can't be measured.*" We must ask: "*What would we need to build to measure it?*"

If resonance and intuition are part of our natural biology, then surely science, which at its best is simply the study of nature, has a responsibility to explore.

The body is transmitting. The signals are real. Let's have the courage to tune in, and create the precision to track what's truly being sent.

It Is Real. We Must Measure It, Methodically and Bravely.

The time for timid language is over.

It *is* real.

Not in the sense of vague belief, but in the sense that it works. Intuition, telepathy, subtle energy — these aren't theoretical musings anymore. They are producing verifiable results, practical outcomes, and patterns too consistent to ignore.

The Truth in the Mirror

The question isn't *"Is it real?"*
The question is, *"Are we ready to study it like it is?"*

Real science doesn't flinch at the unknown. It walks toward it. And yet, for decades, anything to do with energy, intuition, or psychic phenomena has been dismissed — not because it lacked substance, but because it challenged the existing rules of substance itself.

We are no longer in that era.

There are too many examples. Too many case studies. Too many people, from scientists to soldiers, therapists to CEOs, who've had intuitive knowings, telepathic exchanges, near-death experiences, spiritual visits, that changed the course of their work or their lives.

We don't need more dismissal.

We need methodology.

That means:

- Measurable baselines.
- Blind studies with multi-person validation.
- Replicable case reports.
- Frequency tracking over time.
- Gut-brain-heart resonance logs.
- Technological instruments with finer electromagnetic sensitivity.
- AI pattern recognition in speech, sensation, and symbolic data.

It means science stepping up, not just to verify the already accepted, but to pioneer what hasn't yet been understood.

Because we are standing at the edge of a new paradigm:
Where intelligence is not just neural but resonant.

The Truth in the Mirror

Where sensing is not just sensory but vibrational.
Where truth is not just data-driven but field-driven.

And if that sounds impossible, remember this:

We've measured gravity but never seen it.
We use electromagnetic frequencies every day that our bodies cannot feel.
We base entire medical decisions on EEG waves, invisible, intangible brain rhythms.

So why not the gut field?
Why not the energy between beings?
Why not the spiral dance of DNA as it transmits more than just heredity code, but exists to contain and amplify coherence, presence, and signal?

This is not an abandonment of science.

This is its evolution.

And those brave enough to measure it — carefully, ethically, without agenda — may not just unlock the mysteries of intuition. They may also uncover the very architecture of connection itself.

Aligns with Systems Theory, Not Superstition

What we're describing is not fringe mysticism.
It is systems theory in motion.

Energy, intuition, resonance, and subtle sensing are not random phenomena. They are part of a complex, interconnected system, one that modern science is only just beginning to understand.

Systems theory tells us that no part of a system exists in isolation.
Every element influences the whole. Every shift ripples outward.

The Truth in the Mirror

Feedback loops, emergent properties, and nonlinear cause-and-effect are core to how complex systems behave, from ecosystems to organisations, from the brain to the biosphere.

So what if human perception is also a system?
The gut, the brain, the heart, the skin, and the field around us are not separate entities; instead, they are dynamic components of a larger energetic network.

This is not superstition. It is systems awareness.

It's what allows animals to migrate thousands of miles using magnetic fields. It's what lets plants communicate through root networks and fungal threads. It's what creates group flow states, collective fear, mass intuition, and the inexplicable way that one person's energy can change the mood and even the physical temperature of an entire room.

Intuition is not "magic."
It is the human interface with a living, dynamic system of information, one that operates through resonance, coherence, and frequency.

And when we dismiss it as superstition, we don't just deeply insult the people who experience it. We fail to grasp the intelligence of the system itself.

Because the truth is:

- Our gut sends more signals to our brain than the brain sends to the gut.
- Our heart's electromagnetic field radiates far beyond our body.
- Our fascia responds to emotional tension before our conscious mind does.
- Our DNA, coiled and resonant, may be transmitting far more than genetic instructions.

This isn't pseudoscience. It's pre-science.

The Truth in the Mirror

It's what we'll uncover when we finally bring the tools of systems thinking — pattern recognition, feedback mapping, resonance analysis — into the study of energy, intuition, and embodied intelligence.

It's not superstition to say the body knows before the mind.

It's just time we measured how.

Future Health and Learning Depend on This Awareness

We stand at the edge of a new paradigm, one where healing, growth, and even education must incorporate the energetic dimension of life. Because whether we acknowledge it or not, energy is always in the room.

Our health systems are stretched.
Our schools are struggling.
Our culture is saturated with disconnection, dis-ease, and despair.
And underneath all of it, we are missing the signals our bodies are constantly trying to give us.

This is not just about personal growth.
It's about public health.
It's about collective learning.
It's about the future of medicine, education, leadership, and how we relate to ourselves and each other.

When we ignore subtle energy and the intelligence of the intuitive body:

- We miss the early warning signs of emotional collapse and disease, and overlook the real root causes of anxiety, disconnection, and chronic illness.
- We teach children to abandon their own knowing in favour of data that ignores their reality and raise generations out of tune with their own bodies and with each other.

The Truth in the Mirror

- We waste lifetimes chasing cures, when what's needed is reconnection.

But when we *include* subtle energy and embodied awareness:

- We catch dysregulation earlier, before it becomes illness, to prevent disease rather than just treat it.
- We teach children how to listen to their own nervous systems and find safety in themselves.
- We reduce the overreliance on pills, protocols, and external fixes.
- We build systems that honour both science *and* sensitivity, validate neurodiverse experience as responsiveness, not dysfunction.
- We learn to work with life force, not against it, restoring the broken bridge between body and soul — and through it, between science and spirit.

Imagine a healthcare system where gut-brain coherence is measured and mapped.
Where trauma is addressed not just through talk, but through frequency and fascia.
Where emotional intelligence is tracked through heart rate variability, and psychic sensitivity is studied, supported, and understood — not dismissed.

Imagine schools where children are taught breathwork, creativity and movement as core subjects.
Where sensory overwhelm is acknowledged as real.
Where neurodiverse brilliance is seen not as disorder, but as a different kind of antenna — one tuned into signals most of us have learned to suppress.

Imagine workplaces where intuitive leadership is valued.
Where emotional resonance is recognised as a strategic asset.
Where humans are treated as energetic beings, not just economic units.

None of this is fantasy.

The Truth in the Mirror

It is simply what becomes possible when we allow energy to be part of the conversation.

We are not just nudging into a new era — we are standing at the crossroads of survival.

My book, The Intelligent Body, was written as a bridge between instinct and intellect, body and belief, science and subtle sensing. It laid the groundwork for what we are saying here, louder and clearer: the body is not just physical. It is energetic. Intelligent. Aware. And we cannot continue to pretend otherwise.

If we want a future where human beings are well — truly well — we must bring this understanding to the forefront.

This isn't about throwing away science. It's about expanding it. It's about opening the lens so wide that we stop filtering out the parts of life that don't fit our current models.

Because energy does fit.
We just haven't been measuring it properly.

Health must become more than physical resilience. Education must become more than cognitive load. Leadership must become more than performance metrics and big egos.

This is the work. This is the evolution.
And it begins by accepting the truth we've been dancing around for decades:

We are energetic beings, living in an intelligent body.
And we are running out of time to pretend otherwise.

Join the Evolution: A Public, Ethical Invitation

This is not just a book. It is a call.

A call to researchers and readers.

The Truth in the Mirror

To sensitives and scientists.

To those who've always known there was more to life than what we could measure, and to those brave enough to try and measure it anyway.

We are standing at the frontier of a new kind of intelligence, one that doesn't separate the body from the mind, the gut from the heart, the brain from the field of energy that surrounds us all. One that sees intuition not as superstition but as the original language of life.

But we cannot move forward alone. This is your invitation to step into the field.

If you are a physicist, a vet, a medical doctor, a psychologist, a neuroscientist, a psychiatrist, a gut or heart specialist, a systems theorist, a frequency analyst, a biofield researcher, a clair-sensing practitioner, or someone who knows in your bones that our current models are too small, then this book was written for you.

Let us study what has always been sacred.

Let us test what has long been true.

Let us find new language, new tools, new courage.

Let us not be afraid to collaborate across disciplines, across belief systems, across cultures, across paradigms.

Because the body is already waiting.

Because the animals already know.

Because our planet is already calling.

And because the future of human thriving depends on our willingness to evolve ethically, publicly, and together.

This is the invitation.

Not to believe.

But to begin.

The Truth in the Mirror

Chapter 9: Training the Mind-Body Instrument

Making the mysterious teachable.

I've Taught a Vet and a Male Accountant — This is Universal

You might think you have to be mystical to do this work.
You don't.

I've taught a vet.
I've taught a male accountant.
I've taught people from many different backgrounds, many of whom arrived assuming they weren't "intuitive" at all, but they were interested and open-minded enough to try.

What they discovered wasn't magic.
It was that their bodies already knew.

Animal communication and the activation of the psychic or clair senses aren't reserved for special people. They are biological capacities, wired into each of our nervous systems, our electromagnetic fields, our gut-brain axis, and our subtle awareness. The challenge is not learning how to receive. It's remembering how to listen.

And here's something else I've noticed:
Some of the most naturally sensitive people I've worked with have also been those with trauma.

Not because trauma is desirable.
But because it sharpens awareness.

The Truth in the Mirror

When you've had to survive, your body tunes in. You read micro-signals. You sense danger before it arrives. You become hyper-aware of tone, vibration, body language, and space. These same skills, when healed and rebalanced, become incredible tools for subtle communication. The difference is survival lives in fear. Intuition lives in trust.

So part of training the intuitive body is not just about technique. It's about safety. Regulation. Rebuilding trust in your own perception, without flooding your system.

It also means shifting from transmission to reception.
You cannot receive intuitive data while your ego is busy performing.
You cannot hear an animal clearly while your mind is trying to fix, prove, validate, or please.
You must learn to stop. Listen. Let the message come.

That requires something most people are never taught:
How to feel from the body.
Not think you're feeling.
But actually feel.

The brain is not the centre of this work.
It is a magnificent processor, but it is not the source.

Your brain sits in cerebrospinal fluid, it is isolated from the rest of the body.
It cannot feel.
It cannot control.
It must learn to listen.

The body, with its spirals, fields, and instinctive signals, is the source.
The brain must interpret, not override. That's the key.

And it's why this work is so powerful when taught well.

The Truth in the Mirror

Because it's not about performance.
It's about perception.

And perception — clean, clear, congruent perception — can be trained.

You Demystify, Systematize, and Build Confidence

One of the most radical things I do in this work is make it ordinary.

Not diluted. Not reduced. But integrated.
Because the mystical becomes teachable the moment we stop pretending it's mysterious.

There's a method to intuitive development.
There are protocols, practices, and structures that work.
There's a way to approach subtle energy that builds skill without inviting fantasy.

I've taught people who thought they weren't intuitive, and within days, they were accurately reading animals, describing environments, picking up health issues, and translating emotional needs with clarity and calm. Not because they suddenly "became psychic," but because they followed instruction that bypasses the noise of the mind and activates the intelligence of the body.

Intuition isn't a performance.
It's a process.

And like any process, it improves with practice. Some take longer than others to get there, as is often the case with any group of people, whether young or old, who are learning a new skill.
The more you learn to regulate your nervous system, quiet your analytical brain, and attune to the right internal channels, the more accurately you receive. The more trust you build.

That trust is everything. Because without it, people second-guess what they receive.

The Truth in the Mirror

Or worse, they start to perform.
And performance always leads to distortion.

The key is knowing how to teach people to receive data *cleanly*, without inserting their opinions, their projections, or their desire to help. Especially when it comes to animals, the line between empathy and assumption can be razor thin. That's why my method is rigorous. Kind, yes. Encouraging, yes. But also structured.

Because we don't need more drama in the intuitive arts.
We need more integrity.

And that's what builds confidence.
Not just in the reader, but in the people and animals receiving the messages.

You're not guessing.
You're not hoping.
You're receiving.
And that's a skill anyone can learn, if they're willing to slow down, stay present, and let the body lead.

Training Creates Repeatable Outcomes in Animal Readings

This work isn't random.

When taught correctly — with presence, ethics, and grounded methodology — intuitive and telepathic communication becomes repeatable. Verifiable. Trustworthy.

That's not a vague claim. It's something I've witnessed again and again through the trainings I've run, the case studies I've gathered, and the sessions I've either conducted or experienced.

The Truth in the Mirror

When two or more people, training with me and Erin, tune into the same animal, without conferring, without prior knowledge, and deliver matching or highly compatible insights? That's not luck. That's systematized sensitivity.

I've seen it with beginners.
I've seen it with skeptics.
I've seen it with people who thought they were "too left-brained" to ever access this kind of work.

The repeatability of these outcomes is what gives this work credibility. It's also what begins to move it out of the "woo-woo" category and into the realm of teachable, testable human skill.

Just like you can train someone to play the piano, paint a portrait, or perform surgery, you can train someone to receive and read energy.
To tune into a non-verbal being.
To receive clear, sensory telepathic data from a body that isn't theirs.

This doesn't mean everyone will receive in the same way, far from it. Some people feel, others see. Some hear, some simply know. The channels are different. But the results can be astonishingly aligned when the system works.

And the system does work.
Not because it's mine.
But because it honours the intelligence of the body.
It removes the ego. It steadies the nervous system.
And it builds in boundaries, discernment, and repeatable practice.

If you're looking for proof, don't just ask for it.
Train for it.

That's where the real confidence begins.

The Truth in the Mirror

The Difference Between Learning and Performing

One of the most common blocks to intuitive development isn't ability — it's performance pressure.

People don't realise this work is not about "getting it right" on command, or "proving" they can communicate, or "fearing" sounding silly or weird. It's about learning to listen differently. To feel without rushing to explain. To receive without immediately having to prove.

We are so conditioned to perform, to produce answers, make things logical, speak quickly and confidently. But the intuitive body doesn't work like that. It whispers. It shows you fragments. It waits for safety and spaciousness before revealing its truth.

If you try to perform this work before you've learned how to feel, you'll either freeze up or default to your brain's guesses. And guesses are not intuition — they are noise.

I've watched talented people lose access to their natural insight because they were too focused on "doing it well" or needing to be "seen to be clever". I've also watched quiet, hesitant learners, people who doubted themselves deeply, become profoundly accurate once they let go of the fear of failure.

You don't get better by performing. You get better by practicing without pressure.
By tracking your own signals.
By noticing how you feel when a message is clean versus when its ego filtered.

And you learn to discern the difference not by being told, but by experiencing it again and again, in real time, with real beings, under real ethical care.

This work is not a talent show.
It's a sacred conversation.
You don't need to be dramatic, or mystical, or wildly confident to be good at it.

The Truth in the Mirror

You need to be *present*.

You need to be *clean*.

And you need to know the difference between performance and truth.

That's what makes this skill transferable.

That's what makes it trustworthy with training.

Building Trust Without Bypassing Rigour

Trust is essential in intuitive work, but blind trust is dangerous.

You must learn to trust your intuitive perceptions. But that trust must be earned through awareness, repetition, and discernment. Not through ego. Not through spiritual bypassing. And certainly not through performance.

This is where a lot of intuitive teaching has gone astray.

People are encouraged to "just believe," to "trust what they get," or to "share everything that comes through." But sharing everything, without question, without reflection, without cross-checking, isn't intuitive mastery. It's recklessness.

True trust is built the same way in intuition as it is anywhere else: through consistency, accountability, case studies, validation, results, and honest self-reflection.

In my own journey, I've had to build that trust in myself session by session. I didn't begin by believing everything I received. I tested it. I logged it. I validated it with owners and animals, with clients and card readings. I noticed when I was right. I noticed when I wasn't. And I learned the difference between a clean signal and a distorted one, an accurate interpretation and an assumptive one.

That learning never ends. It evolves with every reading.

The Truth in the Mirror

One of the most dangerous traps in this work is the assumption that if something "feels strong," it must be true. But strong signals can come from trauma just as easily as from truth. This is why rigorous intuitive training matters. It's why ethical frameworks and self-awareness are non-negotiable.

When we teach animal communication, we don't just teach people to receive impressions. We teach them to slow down. To clear their minds. To cross-validate. To listen to their own body's responses, not their stories. We show them how to distinguish between ego and insight, between their own needs and the animal's message.

We track the signals. We test the hits. We review what was validated. We don't skip the part where something might be wrong.

That's not insecurity — that's integrity.

And in doing so, confidence grows, not as blind faith, but as embodied knowing. Confidence that holds, even when others question. Confidence that doesn't need to be flashy. That doesn't perform.

Intuitive development is not about *believing* everything.
It's about *refining* what you receive.
It's about knowing how to listen, without rushing to speak.
It's about respecting the message enough to test it.
And respecting the other enough to not distort it.

Some people believe that spiritual sensitivity should liberate them from structure. I believe the opposite. The more sensitive you become, the more self-discipline you need.

Because when you're truly open, you're also deeply vulnerable to projection, to pressure, to praise. And if you don't have a grounded practice that includes humility, ethics, and feedback, then you're not reading for anyone's benefit but your own.

The Truth in the Mirror

Real trust in this work — the kind that lasts — comes from rigour, not bypass.

You don't have to perform.

You must practice. With heart. With structure. And with respect for the energy that moves through you and the intelligent energy that chooses to trust you.

What Makes a Communicator or Intuitive Energy Reader Credible, Grounded, and Ethical

Intuitive and telepathic work is still too often dismissed as fantasy, or worse, misused as manipulation, and so credibility is everything.

A true communicator or energy reader does not posture, perform, or predict for power.
They tune in.
They listen.
They translate.

And they know what they're doing, and why.

This is not about titles. It's not about how many followers you have or how "spiritual" you sound. It's about what you bring to the field, energetically, ethically, practically, with full presence and aligned truth.

A grounded and credible intuitive energy reader holds a few clear qualities:

They are self-aware. They know their triggers. They've done enough inner work to recognise when a message might be coloured by personal history, projection, or bias. They've learned to pause before speaking and to be honest about uncertainty.

They stay in their lane. A communicator is not a vet, a therapist, or a saviour. A credible reader knows their role, to report what's received, not to interpret beyond

The Truth in the Mirror

their remit. They refer out when needed. They don't diagnose. They don't pretend to know what they don't.

They ask for feedback — and listen to it. They don't need to be right. They're curious about what was accurate, and what wasn't. They understand that validation is how credibility is built, not avoided.

They are not addicted to being special. The intuitive arts attract people who long to feel unique. But the real gift is being able to connect, not to prove. Those who are driven by the need to be "the only one who knows" will always distort the signal. A true reader doesn't need to be the hero. They let the message shine.

They protect the energy — theirs, and yours. Credible readers know how to close space. How to prepare, cleanse, and clear. They understand energy hygiene and hold boundaries that honour the work. You won't find them leaking, venting, or over-exposing others. They respect the sacred.

They do the work — and don't flaunt it. Grounded readers aren't constantly proving themselves. Their work speaks through the results, the shifts, and the clarity it brings. They know that real integrity doesn't need to shout.

They know that the energy doesn't belong to them. It flows through, not from. It is sacred, not staged. The honour is in holding it well, not hoarding it for self-importance.

They remain teachable. No matter how many validations or breakthroughs they've had, credible communicators stay open to learning. The energy field is alive. We are all still discovering. Humility is essential.

When we train people in this work, we're not interested in whether someone is "gifted." We're interested in whether they are willing to listen deeply, humbly, and honestly.

Because the gift isn't the ability to receive.
The gift is what you do with it.

The Truth in the Mirror

A credible communicator or energy intuitive is one who knows how to listen without inserting themselves.

To speak with clarity but never with force.

To channel information without hijacking it.

To honour what is unseen — with precision, integrity, and care.

That is the kind of energy reader the world needs now.

The Truth in the Mirror

Conclusion: The Door is Open

The next step is yours.

The Cage Door Is Open — But You Must Walk Through

There is a point on every intuitive journey where you realise you've been waiting for something.
Permission.
Proof.
A teacher. A sign.
Something to validate what your body has always known: that there is more to life — and to you — than the surface shows.

But here's the truth:
There's a moment every sensitive soul reaches. A moment when you realise, the bars that held you were never locked.

The waiting is over.
The cage door is open.
And the only thing left is to choose.

You can stay where it's familiar. Where logic reigns. Where the mind keeps score and the world claps for your performance.
Or you can step through into something quieter, deeper, and infinitely more true.

Because this journey, the one of psychic awakening, intuitive sensitivity, energetic intelligence, sensory development, telepathic connectivity, is not a fantasy. It's not a superpower. It's not reserved for the gifted or the chosen.

It's a remembering.
A reclamation.

The Truth in the Mirror

A return to the intelligence you were born with, the one that lives not in your credentials or your calendar, but in your physical system, in your cells, your gut, your heart, your knowing.

If you've read this far, it's because something inside you already recognises the truth of what's being said.
Something remembers.
You've felt energy.
You've sensed things before they happened.
You've known things that couldn't be known through logic alone.
You've ignored those knowings, perhaps, or dismissed them as coincidence.
But they were never accidental.
They were invitations.
Each one a crack of light through the bars.

And now you see: those bars were never real.
They were belief systems.
Inherited narratives.
Internalised noise that told you it wasn't safe to trust yourself, that sensitivity made you weak, that logic was king, that anything you couldn't measure wasn't real.

But here's the thing about belief:
It's not truth.
It's habit.

And habits can be unlearned.
Stories can be rewritten.
Doors can be walked through.

You may feel the call quietly, like a pulse in your chest. Or you may feel it thunder through your life, disrupting everything that once made sense. Either way, you are not here by accident.

The Truth in the Mirror

You are here because it is time.
Time to stop performing.
Time to stop pretending you don't know what you know.
Time to stop waiting for someone else to tell you that your inner world is valid.

You do not need more credentials.
You need more courage.
You need trust, not in systems, not in saviours, but in your own felt sense of truth.

This is not about becoming something new.
It's about becoming fully yourself.
It's about reclaiming the parts of you that were silenced, shamed, or shut down.
The parts that felt too much, saw too much, *knew* too much, and were told to stop.

No more.

The door is open.
The world is noisy.
And you are needed, whole, congruent, and awake.

Walk through.
We're waiting.

Reframing Psychic Ability as Innate Human Function

We've been taught that psychic ability is something strange.
Fringe. Unreliable. Special.
This framing is a distortion, a product of centuries of suppression, ridicule, and misunderstanding.

Psychic ability is not some glitter-dusted gift reserved for the few.
It's not about being magical.
It's about being human.
A fully, deeply, attuned human.

The Truth in the Mirror

When we speak of "psychic" senses, we're really speaking of biological capacities that have simply been forgotten, underused, dismissed, discouraged, or overridden by modern life.

The clairs — clairvoyance, clairsentience, claircognizance and their kin — are not fantasies. They are subtle forms of perception felt through a tangible sensory system.

And they live in all of us.

These senses are part of your design.

They whisper through the gut when a choice feels off.

They ripple across the skin when you sense someone's true emotion.

They rise in dreams, images, flashes, or instant knowings, the ones you often dismiss or explain away because they don't follow the logic your brain has been trained to trust.

Just because something isn't measurable by traditional instruments doesn't mean it's not real.

We once believed germs didn't exist because we couldn't see them.

We once believed the heart was only a pump, not the generator of the body's largest electromagnetic field.

We once believed the world was flat.

It is time, long past time, to stop confusing the limits of our current tools with the limits of truth.

Because the real issue is not whether psychic ability exists.

It's whether we are willing to remember it.

To reframe it.

To reawaken to it as the natural birthright of every being on this planet, as integral to our perception as sight, sound, or touch.

And once you do, everything begins to shift.

You stop reaching outside yourself for every answer.

You stop asking others to tell you who you are.

The Truth in the Mirror

You become calmer, more certain of yourself, and you start listening. Deeply.
And in that listening, you reconnect with the intelligence that has always been yours.

Psychic ability isn't extra, it's intrinsic.
It's essential.
It's just been buried under noise, doubt, fear, and the false idea that "rational" means "real."

But this is a new time.
A new chapter.
And a new invitation — to return to the wholeness that has always been waiting for you.

This Is Not About Being Special — It's About Being Willing

One of the greatest misunderstandings about psychic ability is that it marks someone as "special."
Elevated. Chosen. Different.
But real psychic awareness is not about that.
It's not about performance, spectacle, or superiority.

It's about presence.
And permission.

The people who access these abilities most clearly are rarely the loudest in the room.
They're often the ones who've been holding this secret quietly for years, sometimes decades, terrified of what might happen if they spoke.

Because for centuries, psychic perception has been treated as dangerous.
Women were burned for it.

The Truth in the Mirror

Men were disowned.
Children were shamed, medicated, and silenced.

Religious dogma declared it evil.
Scientific fundamentalism declared it impossible.
Social norms deemed it embarrassing, even insane.

So of course, people have hidden.
Of course it has felt unsafe to say: "*I knew something before it happened,*" "*I felt a presence,*" "*I saw the truth in someone's body before they did.*"
This fear is not weakness — it's conditioning.
Generational. Cultural. Internalised.
And it lives deep in our cells, shaping our instincts, shutting our throats.

We have inherited a taboo around knowing.

But the truth is this: we are meant to know.
Not through external approval or institutional validation, but through the wild, wise system of perception built into our very being.

The cost of denying this knowing is not just spiritual.
It's physical.
Emotions stuck in the body become symptoms.
Repression becomes anxiety.
Silencing becomes illness.

We are getting sick as a species, not just because of our diets or our stress or our screens, but because we have forgotten how to listen to ourselves. We have severed the lines of communication between our bodies, our energy, and our souls.

Healing begins when we reclaim those lines.
When we say: "*I don't have to wait for permission to be what I am.*"
When we choose to trust the signals that arise from within.

The Truth in the Mirror

This is not a mystical gift.
This is your human being design.
Your animal body's inheritance.
Your soul's right to sense and respond to life.

To reclaim it is not to become better than others.
It is to become whole again.
And to remember that you are not just a mind with a body.
You are a being with an intelligence that runs far deeper than thought.

So no, this is not about being special.
It's about being willing.

Willing to see.
Willing to feel.
Willing to listen — and to finally believe what your body has been whispering all along.

Integrity Is Non-Negotiable in Intuitive Work

When we begin to truly honour intuitive ability as a natural, teachable human function — not a parlour trick, not a gimmick — something else becomes clear:

This work requires integrity. Always.

Because the moment we open the channel to receive subtle information, we are also holding something sacred:
Another being's truth.
Another being's energy.
Another being's trust.

That is never something to take lightly.

Intuitive reading — whether for animals, people, land, or future timelines — is not about impressing others with flashes of insight.

The Truth in the Mirror

It's about responsibility.

It's about accuracy.

It's about the humility to say "I don't know" when the signal isn't clear.

And it's about never using this work to manipulate, frighten, or impose your will on others.

It is not ethical to tell someone their pet is dying without medical support to back it up.

It is not ethical to project personal opinions into a reading and dress them up as guidance.

It is not ethical to pretend you're receiving intuitive downloads when what you're actually doing is guessing or extrapolating from logic.

This work requires rigor.

It requires groundedness.

And it requires the maturity to know when you are fit to read, and when you are not.

Because this is energy work.

And if you are not energetically clean — if you are tired, bitter, anxious, fearful, or projecting — then what you receive will be distorted.

No matter how gifted you are.

It's why I do not read for people in person. It scrambles my frequency.

It's why I don't read when I'm emotionally off-centre.

And it's why I teach every student, no matter their background, to cleanse, ground, and check their ego at the door.

This is not for show.

This is service.

And anyone who steps into this field must be prepared to treat it as such.

The Truth in the Mirror

The good news?
When you do, the work becomes exquisite.
The field becomes clear.
And the information that comes through is not only accurate — it's sacred.

You become a steward.
Not a performer.

You become a channel.
Not a controller.

And that, more than any psychic skill or spiritual title, is what makes you a communicator worth listening to.

We Are Ready for This as a Species

It might not always seem like it.
Turn on the news and you'll see chaos, conflict, and confusion, a world gripped by division, driven by ego, addicted to certainty and performance.

But underneath the noise, something is shifting.

More people than ever are questioning the systems they were taught to trust.
More are listening to their bodies instead of numbing them.
More are sensing the truth before it's spoken.
More are waking up with dreams they can't explain.
More are walking into rooms and feeling what's *really* going on.

They might not call it intuition.
They might not feel ready to speak about it.
But it's happening — and not in isolated pockets.

It's happening across industries, across genders, across cultures and continents.
It's happening to scientists and accountants.
To teachers and therapists.

The Truth in the Mirror

To doctors and dancers.
To people with no prior interest in anything "spiritual," now finding themselves flooded with insights they can't logically explain.

And yes, it's scary for many of them.
We are not just dealing with a lack of information.
We are dealing with centuries of suppression.

People with psychic skills, natural, biological, and often trauma-enhanced skills, have been burned, banished, ridiculed, sectioned, and silenced.
They've learned to keep quiet. To hide.
Because even now, there's a deep-rooted fear that to speak of seeing, hearing, or sensing what others cannot... is to be marked, rejected, cast out.

That fear is cellular. Generational.
And it's making us sick.

We are a species living in the chronic trauma of disconnection.
From land.
From animals.
From our bodies.
From our inner knowing.

It's no wonder anxiety is epidemic.
It's no wonder our health systems are overwhelmed.
It's no wonder so many of us feel numb, exhausted, or lost.

We are deeply entwined in a global state of mental, physical, emotional, and spiritual crisis.

But this isn't the end of the story.

It's the turning point.

The Truth in the Mirror

Because when you reclaim your intuitive senses, not as special gifts, but as your natural intelligence, you begin to heal.
Not just emotionally, but energetically.
Your nervous system calms.
Your immune system strengthens.
Your relationships deepen.
And your sense of purpose, real, soul-level purpose, begins to return.

You are no longer trapped in the performance of being human.
You are living as a being.

And from there, a new world becomes possible.

One rooted in sensitivity, not shame.
One led by soul, not ego.
One where intuition is no longer mocked or mystified, but trained, refined, and honoured.

That world begins when you say yes to your own deeper knowing.

And it continues when we build the structures, scientific, educational, and ethical, to support it.

Because we are ready.
Not just to believe in subtle energy.
But to live by it.

An Invitation to the Future of Sensing, Knowing, and Leading

This book is not the end of anything.
It is a beginning.

The Truth in the Mirror

You've walked with me through intuition, energy, body wisdom, subtle science, and the long-overdue reframing of what it means to know.
You've heard the case studies.
You've seen the logic.
You've felt the resonance.

Now comes the part that only you can do.

You must decide what to do with this awareness.

No one can force you to listen more deeply to your body.
No one can train your mind to trust your senses, or your cells to trust each other.
No one can rewire your reality but you.

But the opportunity is here — alive and waiting.

You are being called not just to become aware of subtle energy, but to apply your awareness.
To bring your intuition into the boardroom.
To bring your energy into your relationships.
To bring your knowing into your leadership.
To stop asking for permission, and start embodying your intelligence, fully.

Because the future of sensing, knowing, and leading will not belong to those who perform best in the old system.
It will belong to those who are brave enough to sense beyond it.

This is not about being psychic.
It's about being present.
To what your body already knows.
To what the field is already broadcasting.
To what your soul has never forgotten.

We are not here to "become" intuitive.
We already are.

The Truth in the Mirror

We are here to *remember* — and to reclaim the sovereign right to trust ourselves again.

That's the invitation.
To the scientist.
To the leader.
To the healer.
To the seeker.
To you.

Look clearly.
Feel deeply.
Respond wisely.

Because the mirror is no longer outside you.
It is within you now.

And the truth?

The truth in the mirror has always been yours to claim.

The Truth in the Mirror

About the Author: Heather Ogilvie

Heather is a former elite turnaround strategist turned pioneering intuitive and energy educator. With a background in global business leadership and a gift for translating subtle perception into grounded insight, Heather bridges the worlds of science and soul. Her journey into intuitive sensing began through animal communication and expanded into a radical reclaiming of the body as a living intelligence system.

She is the creator of *The Truth in the Mirror* series, including *Understanding Women*, *The Intelligent Body*, and *The Intuitive Body*, author of *Fast-Track Millionairess*, founder of *Leading Beyond Logic* and founder of *MoneyMinds Scotland SCIO*, a financial empowerment charity for young adults. Heather's work is rooted in personal transformation, rigorous case studies, and a vision for ethical, embodied leadership in a rapidly evolving world.

When not writing, teaching, or listening deeply to the unseen, she can be found painting or exploring the Scottish landscape with her dogs, honouring the land, or dreaming up new ways to restore dignity, sovereignty, and energetic integrity to human life.

The Truth in the Mirror

Further Support

Follow me to discover more about my programs. Visit my website at www.islaywellness.com or scan the QR code below.

To book an animal communication session with authentic and trusted world-class communicators. I highly recommend:

Erin Furman

Annette Norbury

For outstanding pet photography, I highly recommend Harriet's owner, Selina, who is the creator of our stunning cover photo.

Printed in Great Britain
by Amazon